Inequalities

Michael J. Cloud • Byron C. Drachman
Leonid P. Lebedev

Inequalities

With Applications to Engineering

Second Edition

 Springer

Michael J. Cloud
Department of Electrical Engineering
Lawrence Technological University
Southfield, MI, USA

Byron C. Drachman
Department of Mathematics
Michigan State University
East Lansing, MI, USA

Leonid P. Lebedev
Department of Mathematics
National University of Colombia
Bogotá DC, CUNDINAMARCA
Colombia

ISBN 978-3-319-30770-1   ISBN 978-3-319-05311-0 (eBook)
DOI 10.1007/978-3-319-05311-0
Springer Cham Heidelberg New York Dordrecht London

Mathematics Subject Classification (2010): 26Dxx, 00A69, 00A06

Printed on acid-free paper

Springer is part of Springer Science+Business Media (www.springer.com)

To Daniel Lee Pokora—MJC

To Michael Everett Drachman—BCD

Foreword

This book is chock full of interesting, beautiful, and useful mathematics. It should be studied and consulted by engineers who wish to understand the origins, properties, and limitations of mathematical models commonly used, for instance, in computer simulations of engineering reality. The study of inequalities is essential for understanding the approximation methods required for computing in engineering, which ones to use, and how to use them.

You will find this second edition of *Inequalities* full of enlightening exercises. You will better understand how to analyze and improve computer models of engineering designs. Learn by working through the exercises. You will find it is well worth the effort. Try them first. If you get stuck, there are over thirty pages of generous hints to encourage your progress. Set a reasonable pace for yourself as with any exercise program, and you will get stronger and the exercises will seem easier as you go along.

I am pleased that the authors decided to add a chapter, in this new edition, introducing interval analysis to the reader. It concerns methods for computing intervals enclosing exact results, that is to say simultaneous upper and lower bounds on solutions of equations. This, in spite of errors due to finite approximation of limiting processes, roundoff errors, and uncertainty in input values of measured quantities.

Worthington, Ohio   Ramon E. Moore
January 2014

Preface

One might wonder why it is necessary to study inequalities. Many applied science and engineering problems can be pursued without their explicit mention. Nevertheless, a facility with inequalities is required to understand much of mathematics at intermediate and higher levels. Inequalities serve a natural purpose of comparison. They can provide indirect routes of reasoning or problem solving when more direct routes seem inconvenient or unavailable. Unfortunately, they are much neglected in the typical Western engineering curricula.

This small guide to inequalities was originally written with engineers and applied scientists in mind. Comments from mathematicians who saw the first edition lead us to hope that some mathematicians will find the applications interesting, and that students of mathematics will also find the book useful. A Japanese translation of the first edition, rendered by Satoshi Kaizu of Ibaraki University, was published by Morikita Shuppan in 2004.

The book is intended to help fill the gap between "college algebra" treatments of inequalities and the treatises that exist in the mathematics literature. Unlike classic books on inequalities, it considers topics such as continuity and max–min problems in some detail; although these are traditional topics in calculus, we wish to emphasize that they are based on the solution of inequalities. Important techniques are reinforced through problems at the end of each chapter, and hints are included to expedite the reader's progress. We review a few topics from calculus, but make no attempt at a thorough review. In order to simplify the discussion, we use hypotheses stronger than necessary in some of the statements or proofs of theorems and in some of the problems. For a review of calculus, we recommend the fine classic by Landau [45]. Among the many good books on analysis, we can recommend Stromberg [86].

Two new chapters were added for the second edition. Chapter 6 presents certain inequalities that play important roles in the study of differential equations and the boundary value problems of mechanics. Chapter 7 offers an introduction to interval analysis, a branch of mathematics that provides (among other things) a powerful way of automating work with inequalities. In addition, Chap. 1 has been restructured and new examples and problems appear throughout the book.

We offer our deepest gratitude to Ramon E. Moore, founder of interval analysis, for contributing the Foreword, providing feedback on Chaps. 1 and 7, furnishing the example on automatic differentiation in Chap. 5, and letting us reproduce an example from [32]. We would like to thank Vivian Hutson, Edward Rothwell, Val Drachman, and Natasha Lebedeva for their kind encouragement, and Beth Lannon-Cloud for her critique of the figures. We owe a substantial debt to our Springer editors Eve Mayer, Marc Strauss, Kaitlin Leach, and Ina Lindemann. We want to thank Glen Anderson, Mavina Vamanamurthy, and Matti Vuorinen for their generous help, in particular for pointing out the importance of l'Hôpital's monotone rule and for suggesting several related problems. Dennis Nyquist offered some beneficial comments on the electrical examples in Chap. 1. Finally, we wish to thank Carl Ganser for many valuable discussions and suggestions and for generously agreeing to read the first edition manuscript.

Okemos, MI, USA   Michael J. Cloud
East Lansing, MI, USA   Byron C. Drachman
Bogotá DC, CUNDINAMARCA, Colombia   Leonid P. Lebedev
January 2014

Contents

Chapter 1
Basic Review and Elementary Facts

1.1 Why Study Inequalities?

Inequalities lie at the heart of mathematical analysis. They appear in the definitions of continuity and limit (and hence in the definitions of the integral and the derivative). They play crucial roles in generalizing the notions of distance and vector magnitude. But many problems of physical interest also rely on simple inequality concepts for their solution. In engineering, it is not always best to think in terms of equality. Let us illustrate this statement with a few examples.

Example 1.1. Suppose a continuous electrical waveform $w(t)$, having finite energy, is bandlimited with absolute bandwidth B Hz. This means (Fig. 1.1) that its spectrum $W(f)$ satisfies $W(f) \equiv 0$ for $|f| > B$ (i.e., for $f > B$ and $f < -B$).

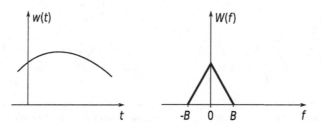

Fig. 1.1 A bandlimited analog waveform $w(t)$ and its hypothetical frequency spectrum $W(f)$. The frequency B is called the absolute bandwidth of $w(t)$

A basic issue in signal processing is how fast we must sample $w(t)$ so that it can be digitized with no loss of fidelity. Straightforward analysis with the Fourier transform shows that if sampling occurs at frequency f_s, then the spectrum of the sampled signal consists of the "baseband" spectrum $W(f)$ replicated every f_s Hz along the frequency axis (Fig. 1.2).

M.J. Cloud et al., *Inequalities: With Applications to Engineering*,
DOI 10.1007/978-3-319-05311-0_1, © Springer International Publishing AG 2014

2 Basic Review and Elementary Facts

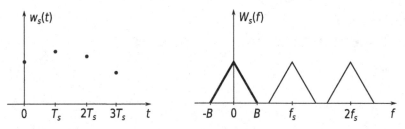

Fig. 1.2 An illustration of the Shannon sampling theorem from signal processing. *Left*: A sequence of samples taken from $w(t)$. The sampling period is T_s seconds and the corresponding sampling rate is $f_s = 1/T_s$ Hz. *Right*: The spectrum $W_s(f)$ of the sampled waveform. The spectrum of the original waveform $w(t)$ is shown in *bold* (cf., the *right side* of Fig. 1.1), while the frequency-translated replicas, produced by the sampling process, are centered at f_s, $2f_s$, and so on. To avoid overlap, we must have $f_s - B \geq B$

To avoid signal corruption through *aliasing* (i.e., through the overlap of adjacent spectral tails on the right side of Fig. 1.2), f_s must be chosen so that $f_s - B \geq B$. The resulting inequality,

$$f_s \geq 2B , \tag{1.1}$$

is the basic content of the Shannon sampling theorem for baseband signals, and $2B$ is the *Nyquist rate* for $w(t)$. If f_s is sufficiently greater than $2B$, we can recover the original signal $w(t)$ from its samples by low-pass filtering with the sort of filter characteristic $H(f)$ indicated on the left side of Fig. 1.3. The filter passes the low-frequency part of the spectrum while rejecting the high-frequency part.

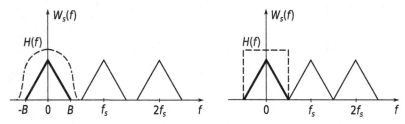

Fig. 1.3 Recovery of $w(t)$ from the sampled signal. *Left*: The situation with oversampling ($f_s > 2B$). The magnitude characteristic $H(f)$ of the "reconstruction filter" (a low-pass filter) is shown *dashed*. *Right*: The situation if sampling is done at the Nyquist rate with $f_s = 2B$. Note that the original waveform $w(t)$ could be recovered from the sampled waveform only by an ideal filter having dashed frequency response $H(f)$

Would it be just as good to state (1.1) as the equality $f_s = 2B$? A sampling process performed *at* the Nyquist rate will result in adjacent spectral copies that touch each other (Fig. 1.3, right). An *ideal* low-pass filter (having brick-wall characteristic indicated by the dashed line in the same part of the figure) will be needed to recover $w(t)$ from the sampled signal. Unfortunately, such filters are not realizable in hardware. This is why oversampling per the inequality (1.1) is done in practice. In this situation it is necessary to think in terms of an inequality.   □

This book is not about signal processing and communication theory, but as those subjects rely heavily on inequalities we provide another example.

Example 1.2. There is a lower bound on the bandwidth B_T of a digital signal:

$$B_T \geq D/2 ,\qquad(1.2)$$

where D is the baud or symbol rate. This is a consequence of the dimensionality theorem of signal theory [16]. Equality would be attained in (1.2) if and only if signaling were done with sinc (i.e., $\sin t/t$ type) pulses—a highly improbable situation in practice. However, as a rule we will pay attention to the circumstances (if any) under which equality holds in a result involving the relation $\leq$ or $\geq$.   □

Let us consider two more examples from electrical engineering.

Example 1.3. If two electrical coils have self-inductance values L_1 and L_2, then their mutual inductance satisfies

$$M \leq \sqrt{L_1 L_2} .\qquad(1.3)$$

This relation is fundamental to the study of inductance. A derivation can be based on basic flux considerations. Figure 1.4 shows a pair of circuits, with circuit 1 carrying electric current I_1.

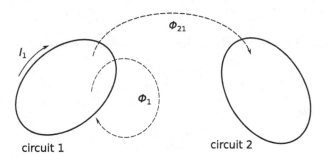

Fig. 1.4 Magnetic flux linkage between electrically isolated circuits. The current I_1 in circuit 1 produces magnetic flux Φ_1 through circuit 1 and magnetic flux Φ_{21} through circuit 2. Since all flux lines under consideration pass through circuit 1 but not necessarily through circuit 2, we have $\Phi_{21} \leq \Phi_1$. The circuits may be tightly wound coils having N_1 and N_2 turns of wire, respectively (not shown)

Let Φ_1 be the resulting magnetic flux passing through circuit 1 (assumed to be the same for all N_1 turns of circuit 1), and let Φ_{21} be the flux through circuit 2. Then certainly

$$\Phi_{21} \leq \Phi_1 ,$$

and multiplication of both sides by the positive quantity N_2/I_1 gives

$$\frac{N_2 \Phi_{21}}{I_1} \leq \frac{N_2}{N_1} \cdot \frac{N_1 \Phi_1}{I_1} .$$

On the left we have the definition of the mutual inductance coefficient M_{21}, while on the right we have the ratio N_2/N_1 multiplying the self-inductance value L_1. Hence

$$M_{21} \leq \frac{N_2}{N_1} L_1 .$$

Interchanging subscripts (i.e., performing the experiment the opposite way, passing a current through coil 2 instead) we get

$$M_{12} \leq \frac{N_1}{N_2} L_2 .$$

Multiplying these last two inequalities, using the reciprocity relation $M = M_{12} = M_{21}$, and taking the square root of both sides, we obtain (1.3).

Another derivation of (1.3) is based on energy considerations [26]. When the two coils carry currents I_1 and I_2, respectively, the total stored energy is given by

$$W = \tfrac{1}{2} L_1 I_1^2 + \tfrac{1}{2} L_2 I_2^2 + M I_1 I_2 .$$

This quantity is nonnegative. By adding and subtracting a term $M^2 I_2^2 / 2 L_1$ on the right-hand side, we can write

$$W = \frac{1}{2} L_1 \left(I_1 + \frac{M}{L_1} I_2 \right)^2 + \frac{1}{2} \left(L_2 - \frac{M^2}{L_1} \right) I_2^2 .$$

In particular, we have $W \geq 0$ when I_2 happens to have the value

$$I_2 = -\frac{L_1 I_1}{M} ;$$

hence we must have

$$L_2 - \frac{M^2}{L_1} \geq 0 ,$$

and this also implies (1.3). $\square$

Example 1.4. Consider the series RLC circuit shown in Fig. 1.5. The voltage source $v_s(t)$ is time-harmonic (sinusoidal) at angular frequency ω. We shall take the loop current $i(t)$ as the output quantity. The capacitor blocks current from flowing at sufficiently low frequencies, while the inductor chokes off current at sufficiently high frequencies. The amplitude response (plot of transfer function magnitude $|H(\omega)|$ vs. ω) exhibits a peak at frequency

$$\omega_0 = 1/\sqrt{LC} .$$

At this *resonant frequency* the impedance seen by the voltage source is purely real and equal to the resistance value R so that $|H(\omega_0)| = 1/R$. The other two displayed frequencies ω_l and ω_u are, respectively, the lower and upper half-power frequencies.

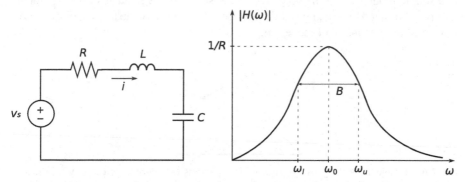

Fig. 1.5 Series electrical resonance. *Left*: A series RLC circuit excited by a time-harmonic (AC) voltage source $v_s(t)$ set at frequency ω. The current $i(t)$ is also time-harmonic at frequency ω. *Right*: Amplitude response of the circuit

The resonant bandwidth and quality factor of the network are given by

$$B = \omega_u - \omega_l \quad \text{and} \quad Q = \omega_0/B .$$

Electrical engineers use B to measure the absolute width of the response curve (in rad/s) and Q to measure sharpness-of-peak (width of the curve relative to ω_0). A bit of circuit analysis shows that

$$\omega_l = \omega_0 \left[\sqrt{1 + \left(\frac{1}{2Q}\right)^2} - \frac{1}{2Q} \right] , \qquad \omega_u = \omega_0 \left[\sqrt{1 + \left(\frac{1}{2Q}\right)^2} + \frac{1}{2Q} \right] ,$$

and by multiplying these equations we find that $\omega_l \omega_u = \omega_0^2$. Hence the resonant frequency lies at the geometric mean of the two half-power frequencies:

$$\omega_0 = \sqrt{\omega_l \omega_u} .$$

A famous inequality states that the geometric mean of a set of positive real numbers cannot exceed the arithmetic mean of those numbers, with equality holding if and only if the numbers are all equal. In the present example this implies

$$\omega_0 < \tfrac{1}{2}(\omega_l + \omega_u) .$$

The inequality of the means has many applications. A more general version is presented as Theorem 3.3 in Sect. 3.4.   □

Now we present a biomechanical example. Although elementary, it indicates the value of being able to think in terms of inequality.

Example 1.5. Consider a person walking across flat, nonslippery terrain. One leg swings forward pendulum-like while the other foot is planted firmly on the ground. In a simple model [2] for the body during the stride, the leg attached to the planted foot is represented by a straight, rigid member of length L, while the rest of the body is a point mass m on top of the leg (Fig. 1.6).

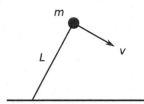

Fig. 1.6 Example on walking. The point mass m represents the entire body exclusive of the leg currently planted on the ground. The inequality implied by this model is $v^2/L \le g$, which yields relation (1.4)

The point mass describes a circular arc at some tangential velocity v and having centripetal acceleration v^2/L. This value certainly cannot exceed the free-fall acceleration g, and we have

$$v \le \sqrt{gL}. \tag{1.4}$$

We can use this to estimate the speed at which a transition from the walking gait to a running gait must occur for an average person with $L \approx 0.9\,\mathrm{m}$; since $g = 9.8\,\mathrm{m/s^2}$, the model suggests that the person must run to exceed a speed of 3 m/s. □

We return to electrical engineering for our final examples of this section.

Example 1.6. The parallel combination of a set of electrical resistors $R_1, \ldots, R_m$ gives an equivalent resistance R_e that cannot exceed any of the individual resistances in the set (Fig. 1.7). Electrical engineering students are taught to remember this fact as a way of checking their calculations. To see why it holds, we can write

$$R_e^{-1} = R_1^{-1} + \cdots + R_m^{-1} \ge R_n^{-1} \qquad (n = 1, \ldots, m) \tag{1.5}$$

and take reciprocals to get $R_e \le R_n$ for any $n = 1, \ldots, m$.

An easy analysis with inequalities shows that electric current divides among parallel resistors in such a way that power dissipation is minimized. Suppose resistors $R_1, \ldots, R_m$ are connected in parallel, let i be the total current entering the network, and let i_n be the current through R_n. The resistors all share the same voltage v (by definition of two-terminal electrical elements in parallel). The power dissipated in R_n is $i_n^2 R_n$, and the total power dissipated is given by

$$P = \sum_{n=1}^{m} i_n^2 R_n.$$

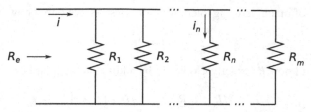

Fig. 1.7 A set of m electrical resistors connected in parallel. The equivalent resistance R_e is less than the smallest of the individual resistances $R_1, \ldots, R_m$. Moreover, the total current i entering the combination distributes itself in such a way that the total dissipated power is minimized

Consider now what would happen if the total current i were distributed in some other way. Letting the current through R_n be $i_n + \delta_n$ instead, the constraint

$$\sum_{n=1}^{m} (i_n + \delta_n) = i$$

implies that the δ_n sum to 0. If P' is the new dissipated power, then

$$P' - P = \sum_{n=1}^{m} (i_n + \delta_n)^2 R_n - \sum_{n=1}^{m} i_n^2 R_n = 2v \sum_{n=1}^{m} \delta_n + \sum_{n=1}^{m} \delta_n^2 R_n .$$

Hence

$$P' - P = \sum_{n=1}^{m} \delta_n^2 R_n \geq 0$$

and we conclude that $P' \geq P$. $\qquad\qquad\qquad\square$

Example 1.7. Inequalities can track the propagation of uncertainty through an algebraic formula. Let us take (1.5) with $m = 2$, simplifying the notation by setting $R_1 = R$ and $R_2 = r$:

$$1/R_e = 1/R + 1/r . \qquad\qquad (1.6)$$

The network under consideration is shown in Fig. 1.8.

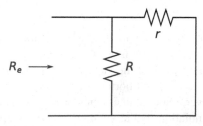

Fig. 1.8 The case of two parallel resistors. The resistance values R and r are known only within certain manufacturing tolerances

If, because of manufacturing tolerances for resistors, we only know that

$$\underline{R} \le R \le \overline{R} \quad \text{and} \quad \underline{r} \le r \le \overline{r} \, ,$$

then we cannot know R_e exactly. However, by adding the inequalities

$$1/\overline{R} \le 1/R \le 1/\underline{R} \, , \qquad 1/\overline{r} \le 1/r \le 1/\underline{r} \, ,$$

and taking reciprocals, we can bound R_e as

$$(1/\underline{R} + 1/\underline{r})^{-1} \le R_e \le (1/\overline{R} + 1/\overline{r})^{-1} \, . \tag{1.7}$$

As a numerical example, suppose $R = 1000 \pm 10\%$ and $r = 100 \pm 1\%$ (resistor values are typically specified in this way, as nominal values with percent tolerances). Then $\underline{R} = 900, \overline{R} = 1{,}100, \underline{r} = 99, \overline{r} = 101$, and we have

$$89.189 \ldots \le R_e \le 92.506 \ldots \, .$$

What if we are presented with a formula far more complicated than (1.6)? This question is addressed by *interval analysis*, a branch of mathematics that offers a way of working with closed intervals as elements of a new number system. The connection between inequalities and closed intervals is that we have $x \in [a, b]$ if and only if $a \le x \le b$. The interval number system extends the real number system, because every real number x can be regarded as a degenerate interval $[x, x]$. However, in interval analysis we can work directly with intervals such as $[\underline{R}, \overline{R}]$ and $[\underline{r}, \overline{r}]$, which contain the respective real values R and r. The system provides ways to add intervals, take reciprocals of intervals, and so on, and the results of these operations are intervals. In our present example, interval computations would yield a set membership statement

$$R_e \in [(1/\underline{R} + 1/\underline{r})^{-1} \, , \, (1/\overline{R} + 1/\overline{r})^{-1}]$$

corresponding to the inequality (1.7). Hence inequality manipulations such as those exhibited above can be automated. Moreover, this can be done in such a way that rigorous enclosures of desired quantities are obtained despite the round off error inherent in finite representation computer arithmetic. We will provide a brief introduction to interval analysis in Chap. 7.   □

In elementary mathematics, inequalities are relations—between numbers or expressions—specified by the symbols $<$, $\le$, $>$, or $\ge$. The task of solving an inequality entails finding the values of the variables that appear in the inequality, for which the inequality holds. A good portion of the present book is devoted to this type of question. However, many mathematics problems lack explicit reference to inequalities in their formulations but are nonetheless of this type. For example, the minimum points of a function f are actually solutions x_0 of the inequality

$$f(x_0) \le f(x) \, . \tag{1.8}$$

We may seek a global solution to (1.8), valid for all x from the domain of f, or a local solution x_0 along with a neighborhood N of x_0 such that (1.8) holds whenever $x \in N$. This is an optimization problem. The class of optimization problems extends to the min–max problems for functionals, which are integral expressions that take real values and whose integrands contain some unknown functions. Here the solution again reduces to finding a solution of an inequality analogous to (1.8). There are other problems that involve inequalities. For example, according to the usual ε-δ definition of continuity, to prove that a function f is continuous at a point x_0, we must prove that the solution of the inequality $|f(x) - f(x_0)| < \varepsilon$ contains the interval $|x - x_0| < \delta$ for some $\delta > 0$. When we solve a problem numerically or use an analytical approximation procedure, we are interested in knowing the error in the approximation with respect to the exact solution. This is by no means a complete list of problems which reduce to the solution of inequalities.

The present book considers various inequality problems in the broad sense in which they are met in engineering practice. It does not cover all such problems— only the most frequent ones.

1.2 Quick Review of the Basics

A few set and logic symbols will serve as convenient shorthand:

$\in$	set membership	$\mathbb{N}$	the natural numbers (positive integers)
$\subseteq$	subset containment	$\mathbb{C}$	the complex numbers
$\cup$	set union	$\mathbb{R}$	the real numbers
$\cap$	set intersection	$\mathcal{P}$	a proposition

The set-builder notation $S = \{x \colon \mathcal{P}(x)\}$ specifies S as the set of all elements x such that proposition $\mathcal{P}(x)$ holds. For instance, $\{x \in \mathbb{R} \colon x^2 = 1\}$ is the set of all real solutions of the equation $x^2 = 1$ and is the same as the set $\{-1, 1\}$. We will occasionally use the symbols

$\Longrightarrow$ for logical implication

$\Longleftrightarrow$ for logical equivalence ("if and only if")

The positive real numbers are separated from the negative real numbers by the real number zero, and we say that $a \leq b$ if $b - a$ is positive or zero. Similarly, we introduce the relation $a < b$ and the reverse symbols $\geq$ and $>$. We say that the inequalities $a < b$ and $c < d$ have the *same sense*, while $a < b$ and $c > d$ have *opposite sense*. Inequalities such as $a < b$, where equality is precluded, are sometimes called *strict*. If equality can hold (as in, for instance, $a \leq b$), the inequality is said to be *weak* or *mixed*.

Example 1.8. A closed interval $[a, b]$ is defined as $[a, b] = \{x \in \mathbb{R} \colon a \leq x \leq b\}$. Finite intervals of the types (a, b), $[a, b)$, and $(a, b]$ are defined analogously. An infinite interval of the form $[a, \infty)$ is defined as the set $\{x \in \mathbb{R} \colon x \geq a\}$. $\square$

The basic laws for inequality manipulation are the axioms [19] that distinguish the set $\mathbb{R}$ as an ordered field. The following hold for any $a, b, c \in \mathbb{R}$.

(a) If $a \leq b$ and $b \leq c$, then $a \leq c$.
(b) We have both $a \leq b$ and $b \leq a$ if and only if $a = b$.
(c) We have $a \leq b$ or $b \leq a$.
(d) If $a \leq b$, then $a + c \leq b + c$.
(e) If $0 \leq a$ and $0 \leq b$, then $0 \leq ab$.

Axiom (a) is the *transitive property*. Axiom (b) is helpful when we want to show indirectly that two real numbers a and b are equal. Note that unlike $\mathbb{R}$, the field $\mathbb{C}$ cannot be ordered. However, any inequality established for real numbers can also be applied to the moduli of complex numbers, and vice versa.

The following list of order properties, while not exhaustive, summarizes many important aspects of inequality manipulation. Suppose $a, b, c, d \in \mathbb{R}$.

1. One and only one of the following holds: $a < b$, $a = b$, $a > b$.
2. If $a \leq b$ and $b < c$, then $a < c$.
3. We have $a \leq b$ if and only if $a + c \leq b + c$; $a < b$ if and only if $a + c < b + c$.
4. If $c < 0$ and $a < b$, then $ac > bc$.
5. We have $a \leq b$ if and only if $a - b \leq 0$.
6. If $a \leq b$ and $c \geq 0$, then $ac \leq bc$.
7. If $a \leq 0$ and $b \geq 0$, then $ab \leq 0$; if $a \leq 0$ and $b \leq 0$, then $ab \geq 0$.
8. We have $a^2 \geq 0$; furthermore, if $a \neq 0$, then $a^2 > 0$.
9. We have $a < 0$ if and only if $1/a < 0$; $a > 0$ if and only if $1/a > 0$.
10. We have $0 < a < b$ if and only if $0 < 1/b < 1/a$.
11. If $a > 0$ and $b > 0$, then $a/b > 0$.
12. If $a < b$ and $c < d$, then $a + c < b + d$.
13. If $0 < a < b$ and $0 < c < d$, then $ac < bd$.
14. If $a \leq b$ and $c \leq d$, then $a + c \leq b + d$.
15. If $a \leq b$ and $c < d$, then $a + c < b + d$.
16. If $a > 1$, then $a^2 > a$. If $0 < a < 1$, then $a^2 < a$.
17. For $c \geq 0$ and $0 < a \leq b$, we have $a^c \leq b^c$, with equality if and only if $b = a$ or $c = 0$.
18. If a is less than every positive real number ε, then $a \leq 0$.
19. If $a \geq 0$ and a is less than every positive real number ε, then $a = 0$.

Note that a term can be transposed to the other side of an inequality if its algebraic sign is changed in the process. Inequalities can be added together, and inequalities between positive numbers can be multiplied. However, *inequalities cannot in general be subtracted or divided*. It suffices to note that subtracting the inequality $1 < 2$ from itself would yield the false result $0 < 0$, and dividing it by itself would yield $1 < 1$. Indeed, some entertaining mathematical sophisms (false arguments intended to deceive) do hinge on invalid inequality manipulations (see, e.g., [10]).

Example 1.9. Suppose $a > b > 0$. Then $a - b$ is positive, and by property 17 we have $a^{a-b} > b^{a-b}$ so that $a^a b^b > a^b b^a$. $\square$

Bounded Sets of Real Numbers

Let S be a set of real numbers. If there is a number B such that $s \leq B$ for every $s \in S$, then S is *bounded above* and B is an *upper bound* for S. Of course, a set that is bounded above has many upper bounds. If there exists M such that M is an upper bound for S and no number less than M is an upper bound for S, then M is called the *least upper bound* or *supremum* for S, denoted by $\sup S$. If $\sup S \in S$, we may refer to $\sup S$ as the *maximum* of S, denoted by $\max S$.

Example 1.10. The interval $I_1 = [0, 1]$ is bounded above by 1 (and by any number greater than 1). No number less than 1 is an upper bound for I_1, so $\sup I_1 = 1$. Moreover, $1 \in I_1$ so that $\max I_1 = 1$. The interval $I_2 = [0, 1)$, on the other hand, has no maximum value although $\sup I_2 = 1$. □

A fundamental property of the real numbers is that any nonempty set of real numbers that is bounded above has a supremum.

We can define analogous concepts of *bounded below* and *greatest lower bound* or *infimum*. The infimum of S is symbolized as $\inf S$, and always exists for sets that are bounded below. If $\inf S \in S$, then S has a *minimum* denoted by $\min S$.

Suppose A and B are, respectively, lower and upper bounds for S:

$$A \leq x \leq B \quad \text{for all } x \in S .$$

If there exist numbers A' and B' such that

$$A < A' \leq x \leq B' < B \quad \text{for all } x \in S ,$$

then A' and B' are also bounds for S but are said to be *sharper* than A and B.

Simple Bounds for Sums

Let $a_1, \ldots, a_N$ be real numbers. Summing the N inequalities

$$\min_{1 \leq n \leq N} a_n \leq a_n \leq \max_{1 \leq n \leq N} a_n \quad (n = 1, \ldots, N)$$

we obtain

$$N \cdot \min_{1 \leq n \leq N} a_n \leq \sum_{n=1}^{N} a_n \leq N \cdot \max_{1 \leq n \leq N} a_n . \tag{1.9}$$

Equality holds if and only if the a_i are all equal. If $b_1, \ldots, b_N$ are nonnegative real numbers, we can sum the inequalities

$$b_n \cdot \min_{1 \leq n \leq N} a_n \leq a_n b_n \leq b_n \cdot \max_{1 \leq n \leq N} a_n \quad (n = 1, \ldots, N)$$

and obtain

$$\min_{1 \le n \le N} a_n \cdot \sum_{n=1}^{N} b_n \le \sum_{n=1}^{N} a_n b_n \le \max_{1 \le n \le N} a_n \cdot \sum_{n=1}^{N} b_n . \tag{1.10}$$

The bounds (1.9) and (1.10) will be used often in what follows.

1.3 Triangle Inequality for Real Numbers

The absolute value of a real number x is given by

$$|x| = \begin{cases} x, & x \ge 0, \\ -x, & x < 0. \end{cases} \tag{1.11}$$

We list some useful properties of the absolute value:

1. $|x| \ge 0$, with equality if and only if $x = 0$;
2. $|xy| = |x|\,|y|$ and, if $y \ne 0$, then $|x/y| = |x|/|y|$;
3. for $a > 0$, we have $|x| < a$ if and only if $-a < x < a$;
4. for $a > 0$, we have $|x| > a$ if and only if $x > a$ or $x < -a$;
5. $-|x| \le x \le |x|$;
6. $xy \le |x|\,|y|$;
7. $|x| \le |y|$ if and only if $x^2 \le y^2$.

Example 1.11. Take a real number x_0 and let $\varepsilon > 0$. The set

$$N_\varepsilon(x_0) = \{x \in \mathbb{R} : |x - x_0| < \varepsilon\}$$

is an *ε-neighborhood* of the point x_0 in $\mathbb{R}$.   □

Example 1.12. Let us show that $|x - 3| < 1$ implies $|x + 2|^{-1} < 1/4$. Indeed, if $|x - 3| < 1$, then $-1 < x - 3 < 1$ and in particular $x + 2 > 4$. Hence $|x + 2| > 4$ and we can finish the argument by taking reciprocals.   □

An elementary but important result is

Theorem 1.1 (Triangle Inequality). *For any two real numbers x and y, we have*

$$|x + y| \le |x| + |y| . \tag{1.12}$$

Equality holds if and only if $x = 0$, or $y = 0$, or x and y have the same sign.

Proof. Left to the reader as Problem 1.1.   □

Using mathematical induction, we can extend Theorem 1.1 for application to more than two real variables.

Theorem 1.2. *For any real numbers $x_1, \ldots, x_n$, we have*

$$\left| \sum_{k=1}^{n} x_k \right| \le \sum_{k=1}^{n} |x_k| \,. \tag{1.13}$$

Equality holds if and only if all the nonzero x_k values have the same sign.

Proof. The case $n = 2$ was stated as Theorem 1.1. Now suppose (1.13) holds for the case $n = m$. To show that it holds for the case $n = m + 1$, we write

$$\left| \sum_{k=1}^{m+1} x_k \right| = \left| \sum_{k=1}^{m} x_k + x_{m+1} \right| \le \left| \sum_{k=1}^{m} x_k \right| + |x_{m+1}| \le \sum_{k=1}^{m} |x_k| + |x_{m+1}| = \sum_{k=1}^{m+1} |x_k| \,.$$

We conclude that (1.13) holds for all integers $n \ge 2$.   □

The next result is closely related to the triangle inequality.

Theorem 1.3. *For any two real numbers x and y, we have*

$$\big| |x| - |y| \big| \le |x + y| \,. \tag{1.14}$$

Equality holds if and only if $x = 0$, or $y = 0$, or x and y have opposite signs.

Proof. By Theorem 1.1, we have $|x| = |x + y - y| \le |x + y| + |y|$ or $|x + y| \ge |x| - |y|$. Swapping the roles of x and y, we get a similar inequality as needed.   □

1.4 Simple Inequalities for Real Functions of One Variable

Suprema and Infima of Functions; Bounded Functions

Let f be a real-valued function with domain D, and let S be a nonempty subset of D. The image of S under f is $f(S) = \{f(x) \colon x \in S\}$ and we write

$$\sup_{x \in S} f(x) = \sup f(S) \quad \text{and} \quad \inf_{x \in S} f(x) = \inf f(S) \,.$$

We say that f is

1. *bounded above* on S if there exists M such that $f(x) \le M$ for all $x \in S$;
2. *bounded below* on S if there exists m such that $f(x) \ge m$ for all $x \in S$;
3. *bounded* on S if it is bounded above and below on S.

Example 1.13. Take $f(x) = 1 - e^{-x^2}$ for $x \in \mathbb{R}$. Then f is bounded on $\mathbb{R}$ because $0 \le f(x) \le 1$ for all x. We also have $\min f = 0$ and $\sup f = 1$, but f does not attain its supremum as a maximum value.   □

Quadratic Inequalities

Consider the quadratic polynomial

$$g(x) = ax^2 + 2bx + c \qquad (a \neq 0)$$

with discriminant $\Delta = b^2 - ac$. Completing the square, we have

$$(1/a)g(x) = (x + b/a)^2 - \Delta/a^2 \ .$$

Therefore, $\Delta \leq 0$ implies $(1/a)g(x) \geq 0$ for all x. Conversely, if $(1/a)g(x) \geq 0$ for all x, then in particular letting $x = -b/a$ gives $\Delta \leq 0$. It is clear that

$(1/a)g(x) \geq 0$ for all x if and only if $\Delta \leq 0$;

$(1/a)g(x) > 0$ for all x if and only if $\Delta < 0$.

Geometrically, $g(x)$ is a parabola with roots $(-b \pm \sqrt{\Delta})/a$ by the quadratic formula. If $\Delta \leq 0$, then $g(x)$ does not have two distinct real roots; hence its graph never crosses the real axis, and $g(x)$ has the same sign everywhere. Conversely, if either $g(x) \geq 0$ for all x or $g(x) \leq 0$ for all x, then $\Delta \leq 0$. If $\Delta < 0$, then $g(x)$ has no real roots; hence its graph never touches the real axis, and $g(x)$ is strictly positive or strictly negative. Conversely, if either $g(x) > 0$ for all x or $g(x) < 0$ for all x, then $\Delta < 0$.

Solving Inequalities in One Real Variable

Given an inequality in a real variable x, we may seek the *solution set*: the set of all $x \in \mathbb{R}$ for which the inequality holds. Suppose f is a real-valued continuous function having a finite number of zeros $(c_1, \ldots, c_n)$ over the interval (c_0, c_{n+1}). So $f(c_k) = 0$ for $k = 1, \ldots, n$. Let us solve the inequality

$$f(x) > 0 \tag{1.15}$$

over $[a, b]$. On each subinterval (c_k, c_{k+1}) $(k = 0, \ldots, n)$, the function f maintains its algebraic sign, which can change only at the zeros of f. This prompts us to determine the sign of f at any point in each subinterval; the result is the *sign graph* of f, which displays the domain over which $f(x)$ is positive.

   If f is a polynomial, the sign graph can be constructed without the calculations at the intermediate points. Consider a polynomial

$$P(x) = a_0 x^n + \cdots + a_n \ .$$

By Bézout's theorem, it can be represented as a product of n factors:

$$P(x) = a_0(x - x_1) \cdots (x - x_n) \ , \tag{1.16}$$

where $x_1, \ldots, x_n$ are the zeros of P. When the polynomial coefficients are real numbers, as is the case when we can consider the inequality $P(x) > 0$, any complex root $x_k = \alpha_k + i\beta_k$ of P (where $\alpha_k, \beta_k \in \mathbb{R}$) is paired with the conjugate root $\alpha_k - i\beta_k$, and the corresponding factor

$$(x - \alpha_k - i\beta_k)(x - \alpha_k + i\beta_k) = (x - \alpha_k)^2 + \beta_k^2 > 0$$

does not affect the sign of $P(x)$ at any x. Hence, to draw the sign graph of P, it suffices to consider only the part of the representation (1.16) that contains the real zeros x_k, which can be multiple. Supposing that $a_0 > 0$ then, we consider an inequality equivalent to $P(x) > 0$:

$$Q(x) = \prod_j'(x - x_j)^{m_j} > 0,$$

where m_j is the multiplicity of the root x_j. The prime indicates that the product includes only terms with real roots x_j.

We initiate the sign graph by noting that $Q(x)$ is positive for sufficiently large x. $Q(x)$ can change sign only at the points x_j. A sign change will actually occur at a given x_j only if m_j is odd.

Example 1.14. Given the polynomial

$$P_1(x) = (x - 1 - i)(x - 1 + i)(x - 1)(x - 2)^2(x - 3)^3,$$

it suffices to consider only the portion

$$Q_1(x) = (x - 1)(x - 2)^2(x - 3)^3.$$

For x large enough, say for $x = 100$, we have $Q_1(100) > 0$ and sign changes can occur only at the points $x = 3, 2,$ and 1. Because $x = 2$ has even multiplicity, sign changes in Q_1 (and hence in P_1) occur only at the points $x = 1$ and 3. So we construct the sign graph from right to left as indicated in Fig. 1.9. The solution set of the inequality $P_1 > 0$ is $(-\infty, 1) \cup (3, \infty)$.

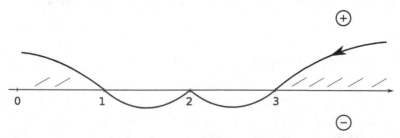

Fig. 1.9 Sign graph for the polynomial P_1. The graph is constructed from right to left as indicated by the *arrow*

It is worth mentioning that the sign graph for P_1 does not change if we move selected factors to the denominator and consider a related inequality for a rational function. For instance, the sign graph for the quotient

$$R(x) = \frac{(x-1)(x-2)^2}{(x-3)^3} > 0$$

is also Fig. 1.9.   □

It is clear that the sign graph technique can also be used to solve weak inequalities of the form $P(x) \geq 0$. A couple of further suggestions are as follows.

1. An inequality $h(x)f(x) > h(x)g(x)$ could be changed to the equivalent form

$$h(x)[f(x) - g(x)] > 0 ,$$

   which holds if and only if $f(x) - g(x) > 0$ and $h(x) > 0$ *or* $f(x) - g(x) < 0$ and $h(x) < 0$.
2. An inequality $f(x)/g(x) > h(x)$ could be changed to the equivalent form

$$\frac{f(x) - h(x)g(x)}{g(x)} > 0$$

   and similar notions applied.

It may be necessary to clear absolute value signs from an inequality before finding the solution set. By property 7 on p. 12, we can replace an inequality of the form $|f(x)| < |g(x)|$ with the equivalent form $f^2(x) < g^2(x)$ and proceed on that basis. More care is required if absolute value signs appear only on one side. Consider an inequality of the form

$$|f(x)| > g(x) .$$

This obviously holds at any point x such that $g(x) < 0$, provided x lies in the domains of both f and g. It also holds at points x where $g(x) \geq 0$ and either $f(x) > g(x)$ or $f(x) < -g(x)$ (in other words, where $g(x) \geq 0$ and $f^2(x) > g^2(x)$). The form

$$|f(x)| < g(x)$$

is equivalent to the pair of requirements $g(x) > 0$ and $-g(x) < f(x) < g(x)$.

Example 1.15. Let us solve $|x-2| > x-1$. The solution set contains any x such that $x - 1 < 0$. When $x - 1 \geq 0$ we can square the inequality to get $(x-2)^2 > (x-1)^2$, which yields the restriction $x < 3/2$. Hence the solution set is $(-\infty, 1) \cup [1, 3/2) = (-\infty, 3/2)$. The given inequality could also be solved graphically as in Fig. 1.10. The value of $|x - 2|$ clearly exceeds the value of $x - 1$ for all $x < 3/2$.   □

Procedures for clearing away radical signs are analogous and are treated in Problem 1.5. Some basic forms involving exponential and logarithmic functions are explored in Problem 1.6. Specific examples appear in Problem 1.7.

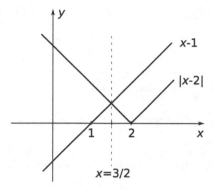

Fig. 1.10 Graphical solution of an inequality involving absolute values (Example 1.15)

1.5 Complex Numbers and Some Complex Functions

With complex numbers involved, inequalities make sense only when written in terms of absolute values. Nonetheless, we collect some elementary facts about complex numbers and certain complex functions, as in the engineering curricula these topics may be presented sporadically. We start with the imaginary unit i, defined as a solution of the equation $x^2 = -1$. Another solution is $-i$. For a complex number $z = x + iy$ where x and y are real, we introduce the real part of z, which is $\mathrm{Re}\, z = x$, the imaginary part $\mathrm{Im}\, z = y$, and the absolute value $|z| = (x^2 + y^2)^{1/2}$. We recall that $z = 0$ if and only if $x = y = 0$. The number z also has the trigonometric form representation $z = r(\cos\phi + i\sin\phi)$. As Fig. 1.11 shows, $x = r\cos\phi$ and $y = r\sin\phi$. Arithmetic actions for complex numbers $z_k = x_k + iy_k$ are defined as follows:

$$z_1 \pm z_2 = (x_1 \pm x_2) + i(y_1 \pm y_2)\,, \qquad z_1 z_2 = (x_1 x_2 - y_1 y_2) + i(x_1 y_2 + x_2 y_1)\,,$$

and

$$\frac{z_1}{z_2} = \frac{x_1 + iy_1}{x_2 + iy_2} = \frac{(x_1 x_2 + y_1 y_2) + i(-x_1 y_2 + y_1 x_2)}{x_2^2 + y_2^2}\,.$$

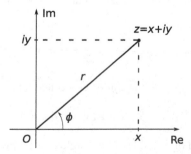

Fig. 1.11 The complex plane. A complex number z is shown along with its real part x, its imaginary part y, its modulus or magnitude r, and its argument or angle ϕ

We now present the triangle inequality for complex numbers.

Theorem 1.4. *Let $z_1, \ldots, z_n$ be nonzero complex numbers. Then*

$$\left| \sum_{i=1}^{n} z_i \right| \leq \sum_{i=1}^{n} |z_i| . \tag{1.17}$$

Equality holds if and only if the z_i all have the same arguments.

Proof. The case $n = 1$ is trivial, so we examine $n = 2$ as the verification step for mathematical induction. Some elementary facts about complex numbers are needed here. For $z \in \mathbb{C}$,

$$\text{Re}[z] = x \leq (x^2 + y^2)^{1/2} = |z| .$$

Similarly, $\text{Im}[z] \leq |z|$. It is also easily shown that

$$|\bar{z}| = |z| , \qquad |z|^2 = z\bar{z} , \qquad \text{Re}[z] = \frac{1}{2}(z + \bar{z}) , \qquad \text{Im}[z] = \frac{1}{2i}(z - \bar{z}) .$$

We use these facts as follows:

$$|z_1 + z_2|^2 = (z_1 + z_2)\overline{(z_1 + z_2)} = |z_1|^2 + |z_2|^2 + 2\,\text{Re}[z_1\bar{z}_2] .$$

However, $2\,\text{Re}[z_1\bar{z}_2] \leq 2|z_1\bar{z}_2| = 2|z_1|\,|z_2|$ so that

$$|z_1 + z_2|^2 \leq |z_1|^2 + |z_2|^2 + 2|z_1|\,|z_2| = (|z_1| + |z_2|)^2 .$$

Taking a square root and noting that both sides are positive, we obtain

$$|z_1 + z_2| \leq |z_1| + |z_2| . \tag{1.18}$$

In general, we have

$$\left| \sum_{i=1}^{n+1} z_i \right| = \left| \sum_{i=1}^{n} z_i + z_{n+1} \right| \leq \left| \sum_{i=1}^{n} z_i \right| + |z_{n+1}| \leq \sum_{i=1}^{n} |z_i| + |z_{n+1}| = \sum_{i=1}^{n+1} |z_i| .$$

The conditions for equality are discussed as part of Problem 4.9.   □

Similarly,

$$|z_1 - z_2|^2 = |z_1|^2 + |z_2|^2 - 2\,\text{Re}[z_1\bar{z}_2] \geq |z_1|^2 + |z_2|^2 - 2|z_1|\,|z_2| = (|z_1| - |z_2|)^2$$

so that $|z_1 - z_2| \geq ||z_1| - |z_2||$, and this can be combined with (1.18) as

$$||z_1| - |z_2|| \leq |z_1 \pm z_2| \leq |z_1| + |z_2| . \tag{1.19}$$

Geometrically, the length of any side of a triangle can neither exceed the sum of the lengths of the two remaining sides, nor fall short of the difference in the lengths of the two remaining sides.

Many analytic functions are extended to complex arguments via their Taylor expansions. Say, for a complex variable z we obtain

$$e^z = \sum_{k=0}^{\infty} \frac{z^k}{k!} , \qquad \cos z = \sum_{k=0}^{\infty} (-1)^k \frac{z^{2k}}{(2k)!} , \qquad \text{etc.}$$

These definitions allow us to verify Euler's formula for real ϕ:

$$e^{i\phi} = \cos \phi + i \sin \phi . \qquad (1.20)$$

As a consequence of (1.20), we obtain the exponential form of z:

$$z = r(\cos \phi + i \sin \phi) = re^{i\phi} .$$

On the other hand we have

$$\cos \phi = \frac{e^{i\phi} + e^{-i\phi}}{2} , \qquad \sin \phi = \frac{e^{i\phi} - e^{-i\phi}}{2i} ,$$

and these relations may be extended to any complex z:

$$\cos z = \frac{e^{iz} + e^{-iz}}{2} , \qquad \sin z = \frac{e^{iz} - e^{-iz}}{2i} .$$

Many familiar properties of the functions e^x, $\sin x$, and $\cos x$ continue to hold when x is replaced by a complex argument z. For instance, we have $e^{z_1} e^{z_2} = e^{z_1 + z_2}$. Other properties are not familiar from elementary mathematics; e^z is periodic with period $2\pi i$, for instance, and $\cos z$ and $\sin z$ become unbounded in the complex plane. The exponential form for z and the periodicity of the trigonometric functions allow us to find n distinct roots of z:

$$z^{1/n} = \left(re^{i(\phi + 2k\pi)} \right)^{1/n} = r^{1/n} \left(\cos \frac{\phi + 2k\pi}{n} + i \sin \frac{\phi + 2k\pi}{n} \right) \quad (k = 0, 1, \ldots, n-1) .$$

Example 1.16. With $z = x + iy$, we have $|\sinh y| \le |\sin z| \le \cosh y$. Indeed

$$\cosh y = \frac{e^y + e^{-y}}{2} , \qquad \sinh y = \frac{e^y - e^{-y}}{2} ,$$

and we can use (1.19) to write

$$\frac{\left| |e^{ix}| |e^{-y}| - |e^{-ix}| |e^y| \right|}{|2i|} \le \left| \frac{e^{i(x+iy)} - e^{-i(x+iy)}}{2i} \right| \le \frac{|e^{ix}| |e^{-y}| + |e^{-ix}| |e^y|}{|2i|}$$

where $|e^{\pm ix}| = |i| = 1$. $\qquad \square$

1.6 Vectors in $\mathbb{R}^n$ and Associated Inequalities

Let us review some concepts regarding vectors. Suppose $\mathbf{x}, \mathbf{y} \in \mathbb{R}^3$. The component representation

$$\mathbf{x} = (x_1, x_2, x_3)$$

implies that three canonical basis vectors $\mathbf{i}_1, \mathbf{i}_2, \mathbf{i}_3$ have been chosen and that

$$\mathbf{x} = x_1 \mathbf{i}_1 + x_2 \mathbf{i}_2 + x_3 \mathbf{i}_3 .$$

The magnitude of $\mathbf{x}$, given by

$$|\mathbf{x}| = (x_1^2 + x_2^2 + x_3^2)^{1/2} , \tag{1.21}$$

yields the length of the segment represented by $\mathbf{x}$ as a position vector. The dot product or inner product

$$\mathbf{x} \cdot \mathbf{y} = x_1 y_1 + x_2 y_2 + x_3 y_3 \tag{1.22}$$

has a clear geometric meaning as the product of the lengths of the vectors $\mathbf{x}, \mathbf{y}$ and the cosine of the included angle. The dot product also has mechanical meaning if we consider $\mathbf{x}$ as a force and $\mathbf{y}$ as a displacement; then $\mathbf{x} \cdot \mathbf{y}$ is the work done by the force $\mathbf{x}$ over the displacement $\mathbf{y}$. Note that the magnitude of a vector can be expressed in terms of the dot product:

$$|\mathbf{x}| = (\mathbf{x} \cdot \mathbf{x})^{1/2} .$$

Analogously, we deal with vectors in $\mathbb{R}^n$ by representing them in the form

$$\mathbf{x} = (x_1, \ldots, x_n) .$$

This means we have introduced some basis $\mathbf{e}_1, \ldots, \mathbf{e}_n$, having exactly n vectors (not necessarily canonical), and

$$\mathbf{x} = x_1 \mathbf{e}_1 + \cdots + x_n \mathbf{e}_n .$$

The equality $\mathbf{x} = \mathbf{0}$ holds if and only if $x_k = 0$ for each k. Expressions (1.21) and (1.22) can be extended for use in $\mathbb{R}^n$, resulting in the Euclidean norm of $\mathbf{x}$ and the Euclidean inner product of $\mathbf{x}$ and $\mathbf{y}$:

$$\|\mathbf{x}\| = \left(\sum_{k=1}^n x_k^2\right)^{1/2} , \qquad \mathbf{x} \cdot \mathbf{y} = \sum_{k=1}^n x_k y_k .$$

These two quantities are related as

$$\|\mathbf{x}\| = (\mathbf{x} \cdot \mathbf{x})^{1/2} .$$

Although the geometric meanings of magnitude and angle disappear when $n > 3$, certain basic properties of the Euclidean norm and its corresponding dot product may be retained and incorporated into other useful expressions. An *inner product* of vectors $\mathbf{x}, \mathbf{y} \in \mathbb{R}^n$ is a function $\langle \mathbf{x}, \mathbf{y} \rangle$ satisfying the following three axioms:

P_1. $\langle \mathbf{x}, \mathbf{x} \rangle \geq 0$, with $\langle \mathbf{x}, \mathbf{x} \rangle = 0$ if and only if $\mathbf{x} = \mathbf{0}$;

P_2. $\langle \mathbf{x}, \mathbf{y} \rangle = \langle \mathbf{y}, \mathbf{x} \rangle$;

P_3.   for any real constants c_1, c_2 and vectors $\mathbf{x}_k, \mathbf{y}$ we have

$$\langle c_1 \mathbf{x}_1 + c_2 \mathbf{x}_2, \mathbf{y} \rangle = c_1 \langle \mathbf{x}_1, \mathbf{y} \rangle + c_2 \langle \mathbf{x}_2, \mathbf{y} \rangle \, .$$

For instance, the expression

$$\langle \mathbf{x}, \mathbf{y} \rangle = x_1 y_1 + \cdots + x_n y_n$$

coincides with the ordinary dot product in the case $n = 3$ for a canonical basis. For a non-canonical basis this is not the case, although properties P_1–P_3 still hold.

An inner product $\langle \mathbf{x}, \mathbf{y} \rangle$ will induce a norm according to the formula

$$\|\mathbf{x}\| = \langle \mathbf{x}, \mathbf{x} \rangle^{1/2} \, . \tag{1.23}$$

In general, a norm $\|\mathbf{x}\|$ is a function satisfying three axioms:

N_1. $\|\mathbf{x}\| \geq 0$, with $\|\mathbf{x}\| = 0$ if and only if $\mathbf{x} = \mathbf{0}$ (*positivity*);

N_2. $\|c\mathbf{x}\| = |c| \, \|\mathbf{x}\|$ for any real number c (*homogeneity*);

N_3. $\|\mathbf{x} + \mathbf{y}\| \leq \|\mathbf{x}\| + \|\mathbf{y}\|$ for any $\mathbf{x}, \mathbf{y} \in \mathbb{R}^n$ (*triangle inequality*).

In Chap. 4 we will introduce abstract spaces, not necessarily finite dimensional, known as inner product spaces and normed spaces. A normed space is a linear space with a norm having a definite value for each element of the space and possessing properties N_1–N_3. An inner product space is a linear space with an inner product defined for any pair of elements and possessing properties P_1–P_3. As mentioned above, an inner product space is a normed space with the *induced* or *natural norm* (1.23). It is clear that in an inner product space, axioms N_1 and N_2 follow from P_1–P_3. Axiom N_3 is verified as follows. First,

$$N_3 \iff \|\mathbf{x} + \mathbf{y}\|^2 \leq \|\mathbf{x}\|^2 + \|\mathbf{y}\|^2 + 2 \|\mathbf{x}\| \, \|\mathbf{y}\|$$

$$\iff \langle \mathbf{x}, \mathbf{x} \rangle + 2\langle \mathbf{x}, \mathbf{y} \rangle + \langle \mathbf{y}, \mathbf{y} \rangle \leq \|\mathbf{x}\|^2 + \|\mathbf{y}\|^2 + 2 \|\mathbf{x}\| \, \|\mathbf{y}\|$$

$$\iff \langle \mathbf{x}, \mathbf{y} \rangle \leq \|\mathbf{x}\| \, \|\mathbf{y}\| \, .$$

Thus, to show that N_3 is valid it is sufficient to prove the following.

Theorem 1.5 (Cauchy–Schwarz Inequality). *For any two vectors* $\mathbf{x}, \mathbf{y} \in \mathbb{R}^n$ *we have*

$$|\langle \mathbf{x}, \mathbf{y} \rangle| \leq \|\mathbf{x}\| \, \|\mathbf{y}\| \, . \tag{1.24}$$

Equality holds if and only if $\mathbf{x} = \mathbf{0}$, *or* $\mathbf{y} = \mathbf{0}$, *or* $\mathbf{x} = \lambda \mathbf{y}$ *for some real* λ.

Proof. If $\mathbf{x} = \mathbf{0}$, or $\mathbf{y} = \mathbf{0}$, or $\mathbf{x} = \lambda\mathbf{y}$ for some $\lambda \in \mathbb{R}$, then equality clearly holds in (1.24). Therefore suppose $\mathbf{x} \neq \mathbf{0}$ and $\mathbf{y} \neq \mathbf{0}$. We will show that (1.24) holds and equality occurs only if $\mathbf{x} = \lambda\mathbf{y}$ for some $\lambda \in \mathbb{R}$. Recall that by P_1 we have

$$\langle \mathbf{x} + \alpha\mathbf{y}, \mathbf{x} + \alpha\mathbf{y}\rangle \geq 0 \ \text{ for all } \alpha \in \mathbb{R}$$

with equality if and only if $\mathbf{x} + \alpha\mathbf{y} = \mathbf{0}$. As $\mathbf{y} \neq \mathbf{0}$, the expression

$$\langle \mathbf{x} + \alpha\mathbf{y}, \mathbf{x} + \alpha\mathbf{y}\rangle = \alpha^2 \|\mathbf{y}\|^2 + 2\alpha\langle\mathbf{x}, \mathbf{y}\rangle + \|\mathbf{x}\|^2 \tag{1.25}$$

is a quadratic polynomial with respect to α which is nonnegative for all α. By the result of Sect. 1.4, its discriminant cannot be positive:

$$\varDelta = \langle\mathbf{x}, \mathbf{y}\rangle^2 - \|\mathbf{x}\|^2 \|\mathbf{y}\|^2 \leq 0 \,,$$

which is equivalent to the needed relation (1.24) for any nonzero $\mathbf{x}, \mathbf{y}$. Equality in (1.24) is equivalent to $\varDelta = 0$; in this case the quadratic (1.25) has the unique real root α_0 for which

$$\alpha_0^2 \|\mathbf{y}\|^2 + 2\alpha_0\langle\mathbf{x}, \mathbf{y}\rangle + \|\mathbf{x}\|^2 = \langle\mathbf{x} + \alpha_0\mathbf{y}, \mathbf{x} + \alpha_0\mathbf{y}\rangle = 0 \,.$$

By P_1, this implies that $\mathbf{x} = \lambda\mathbf{y}$ with $\lambda = -\alpha_0$. $\qquad\qquad\square$

We add a couple of observations:

1. The necessary and sufficient condition for equality in (1.24) can be stated concisely as "the vectors $\mathbf{x}$ and $\mathbf{y}$ are linearly dependent." We say that $\mathbf{x}$ and $\mathbf{y}$ are *linearly dependent* if there exist real constants α and β, not both zero, such that $\alpha\mathbf{x} + \beta\mathbf{y} = \mathbf{0}$.
2. Sometimes the Cauchy–Schwarz inequality is stated in the form

$$\langle\mathbf{x}, \mathbf{y}\rangle \leq \|\mathbf{x}\| \, \|\mathbf{y}\| \,, \tag{1.26}$$

   i.e., without an absolute value on the left-hand side. In this case equality holds if and only if $\mathbf{x} = \mathbf{0}$, or $\mathbf{y} = \mathbf{0}$, or $\mathbf{x} = \lambda\mathbf{y}$ for some *positive* real constant λ. The last condition means that $\mathbf{x}$ and $\mathbf{y}$ are parallel (co-directed with the same sense). Equality holds in (1.24), on the other hand, if $\mathbf{x}$ and $\mathbf{y}$ are parallel or antiparallel.

The Cauchy–Schwarz inequality is important in applied mathematics. We consider special forms of this inequality, along with some of its consequences, in Sect. 3.7.

We should mention that it is possible to introduce infinitely many inner products, along with their corresponding induced norms, on any finite dimensional space. However, it is also possible to introduce norms that cannot be induced by an inner product. One important fact is that all norms on a finite dimensional space are *equivalent*: for any two norms $\|\cdot\|_1$ and $\|\cdot\|_2$ given on the space, there are positive constants c_1 and c_2 such that for any nonzero element $\mathbf{x}$ of the space we have

$$0 < c_1 \leq \frac{\|\mathbf{x}\|_1}{\|\mathbf{x}\|_2} \leq c_2 < \infty \,.$$

This fact—again, pertaining to finite dimensional normed spaces—permits us to employ any convenient norm in order to introduce convergence of a sequence of vectors, as well as derivatives and integrals for vector functions.

1.7 Some Techniques for Establishing Inequalities

The methods needed to deal with mathematical inequalities are numerous and diverse. In this section we outline some common approaches that do not require the use of calculus. Additional methods appear throughout the book.

Reversible Transformations Leading to Known Result

Consider the inequality

$$(n!)^{1/n} < [(n+1)!]^{1/(n+1)} \qquad (n \in \mathbb{N}) . \tag{1.27}$$

We can raise both sides to the $n(n+1)$ power to get

$$(n!)^{n+1} < [(n+1)!]^n$$

and then cancel the factor $(n!)^n$ to get

$$1 \cdot 2 \cdot 3 \cdots \cdots n < \underbrace{(n+1)(n+1)(n+1) \cdots (n+1)}_{n \text{ factors}} . \tag{1.28}$$

It is clear that (1.28) holds for any $n \in \mathbb{N}$. But the fact that we can derive a valid inequality from (1.27) does not, by itself, prove that (1.27) holds.

Indeed, by squaring both sides of the false statement $-1 \geq 1$, we obtain the true statement $1 \geq 1$. This clearly does not prove that $-1 \geq 1$. The problem is that squaring both sides of an inequality is not a reversible transformation unless we are assured that both sides of the inequality are nonnegative.

That being said, the steps leading from (1.27) to (1.28) are actually reversible. With this observation, we have proved (1.27).

Example 1.17. We show that

$$|a + b|^{1/2} \leq |a|^{1/2} + |b|^{1/2} \qquad (a, b \in \mathbb{R}) .$$

Because both sides are nonnegative, we may square both sides to obtain the equivalent statement $|a + b| \leq |a| + |b| + 2|a|^{1/2}|b|^{1/2}$. But this is implied by the triangle inequality, completing the proof. These statements also hold for $a, b \in \mathbb{C}$. ☐

Irreversible Transformations

Sometimes we can create a useful inequality by changing an expression to increase or decrease certain terms. The goal is often to create a simpler expression. We have, for instance,

$$\frac{1}{n^2} < \frac{1}{n(n-1)} \qquad (n > 1) \, .$$

The usefulness of such an observation depends on the circumstances.

Example 1.18. If $n \in \mathbb{N}$ and $n > 1$, then

$$\frac{1}{4n} < \sum_{k=1}^{n} \frac{1}{(n+k)^2} < \frac{1}{n} \, . \tag{1.29}$$

Indeed, an application of (1.9) gives

$$\frac{1}{4n} = n \cdot \min_{1 \le k \le n} \frac{1}{(n+k)^2} < \sum_{k=1}^{n} \frac{1}{(n+k)^2} < n \cdot \max_{1 \le k \le n} \frac{1}{(n+k)^2} = \frac{n}{(n+1)^2} \, .$$

If we replace $n + 1$ by n in the rightmost member, we get (1.29).   □

Substitutions into Known Inequalities

A myriad of results can be obtained as substitution instances of established results. Let us take a simple case. If there is one inequality that can be regarded as the most fundamental, it is arguably that for any real x we have

$$x^2 \ge 0 \tag{1.30}$$

with equality if and only if $x = 0$. Many interesting results can be developed or proved on the basis of (1.30). Replacing x by $x - y$, we have

$$(x - y)^2 \ge 0 \tag{1.31}$$

with equality if and only if $x = y$, and (1.31) can be manipulated into forms such as

$$x^2 + y^2 \ge 2xy \tag{1.32}$$

and

$$2(x^2 + y^2) \ge (x + y)^2 \, . \tag{1.33}$$

Restricting x to nonnegative values, we can use (1.31) to write

$$(x - \sqrt{x})^2 \ge 0 \, ,$$

which in turn yields

$$1 + x \geq 2\sqrt{x} \qquad (x \geq 0).$$

We can easily bring multiple variables into the picture; applying (1.32) three times as

$$x^2 + y^2 \geq 2xy, \qquad y^2 + z^2 \geq 2yz, \qquad z^2 + x^2 \geq 2zx$$

and adding, we obtain

$$x^2 + y^2 + z^2 \geq xy + yz + zx. \tag{1.34}$$

Equality holds if and only if $x = y = z$.

Mathematical Induction

Although we already used mathematical induction to prove (1.13), a brief review of the procedure might be in order. Let $k, p, n \in \mathbb{N}$. A proposition $\mathcal{P}(k)$ holds for all values $k \geq p$ if

1. we prove the validity of $\mathcal{P}(p)$, and
2. supposing the validity of $\mathcal{P}(k)$ for arbitrary $k = n$ ($n \geq p$), we prove that it holds for $k = n + 1$.

Example 1.19. To prove

$$2^k > k + 1 \qquad (k \geq 2) \tag{1.35}$$

we first observe that it holds for $k = 2$. This is the verification step. Now assume it holds for $k = n$:

$$2^n > n + 1.$$

This is the induction hypothesis and may be labeled as $\mathcal{P}(n)$. Multiplying both sides by 2, we get

$$2^{n+1} > 2(n + 1) = 2n + 1 + 1 > (n + 1) + 1$$

and hence $\mathcal{P}(n)$ implies $\mathcal{P}(n + 1)$. □

In a variation called *backward induction*, the principle is that a statement $\mathcal{P}(k)$ holds for all values of k if

1. taking any large N, we find that $\mathcal{P}(k)$ holds for some $k > N$ (which means that it holds for infinitely many k tending to infinity), and
2. the validity of $\mathcal{P}(k)$ implies $\mathcal{P}(k - 1)$ for each k.

See Problem 3.5(a) for an example.

Homogeneity; Constraining or Normalizing the Variables

Many inequalities take a homogeneous form. Such an inequality does not change form when we replace the variables x_k by λx_k with a constant $\lambda > 0$, as the factors of λ all cancel and we are led back to the original inequality. This is the case with the Cauchy–Schwarz inequality, Minkowski inequality, and other classic results covered in Chap. 3. Often homogeneous inequalities are proved by first "normalizing" the variables, in order to simply the expressions involved, and then transitioning to the general case. First we introduce the required notion. Let g be a function of the n variables $x_1, \ldots, x_n$. We say that g is *homogeneous of degree k* if the equation

$$g(\lambda x_1, \ldots, \lambda x_n) = \lambda^k g(x_1, \ldots, x_n)$$

holds for an arbitrary real number λ. If both sides of an inequality

$$g(x_1, \ldots, x_n) \geq h(x_1, \ldots, x_n)$$

are homogeneous functions of degree k, then by setting $f = g - h$ we can write the inequality in the form

$$f(x_1, \ldots, x_n) \geq 0 \tag{1.36}$$

where f is homogeneous of degree k. Assuming the arguments $x_1, \ldots, x_n$ are positive, we can recast (1.36) into the form of a constrained inequality: that is, an inequality along with a constraint equation satisfied by its variables. The constrained inequality may be simpler to prove.

Theorem 1.6. *Let f be homogeneous of degree k. Then the inequality*

$$f(x_1, \ldots, x_n) \geq 0 \ \text{for any positive} \ x_1, \ldots, x_n \tag{1.37}$$

is equivalent to the inequality

$$f(a_1, \ldots, a_n) \geq 0 \tag{1.38}$$

for any positive $a_1, \ldots, a_n$ constrained or normalized in various ways such as

$$a_1 + \cdots + a_n = 1 , \quad a_1^2 + \cdots + a_n^2 = 1 , \quad a_1 \cdots a_n = 1 , \quad or \quad a_n = 1 . \tag{1.39}$$

Proof. By s we denote the left member of the normalization expression. For the first constraint in (1.39), say, we put $s = a_1 + \cdots + a_n$. So assume (1.38) holds along with this constraint, and let $x_1, \ldots, x_n$ be any given set of positive numbers. Put $a_i = x_i/s$ $(i = 1, \ldots, n)$ into (1.38); this gives

$$f(a_1, \ldots, a_n) = f(x_1/s, \ldots, x_n/s) = (1/s)^k f(x_1, \ldots, x_n) \geq 0 ,$$

which yields (1.37). The converse is obvious.   □

Example 1.20. The inequality

$$(a + b)(a^{-1} + b^{-1}) \geq 4 \qquad (a, b > 0) \tag{1.40}$$

is homogeneous of degree zero and, without loss of generality, we may consider it under the restriction that the variables sum to 1. We can manipulate it into the form $(a + b)^2/ab \geq 4$ and assume that $a + b = 1$ in order to reduce it to the simpler inequality $ab \leq 1/4$. This clearly holds for any two numbers that add to unity. □

It is permissible to normalize the problem (1.37) by setting $x_i = 1$ for one particular i. This allows us to decrease the number of variables by one.

Example 1.21. To prove (1.40), we can set $b = 1$ and treat the equivalent inequality

$$(a + 1)(a^{-1} + 1) \geq 4 \qquad (a > 0) .$$

This is easily deduced from the fundamental inequality $(a - 1)^2 \geq 0$. □

See Problem 1.17 for more opportunities to exploit homogeneity.

Ordering of the Variables

If the variables in a set of variables are known to be ordered in a certain way, it may be possible to derive an interesting inequality from that information. Suppose, for instance, we are dealing with four real variables a, b, x, y and it is known that $a \geq b$ and $x \geq y$. Then $a-b$ and $x-y$ are both nonnegative and we have $0 \leq (a-b)(x-y) = ax + by - bx - ay$, which gives us

$$ax + by \geq ay + bx .$$

Some standard inequalities can be developed along similar lines.

Example 1.22 ([49]). Take $a, b \in \mathbb{R}$ with $a \neq b$, and $m, n \in \mathbb{N}$. The numbers $a^m - b^m$ and $a^n - b^n$ are either both positive (when $a > b$) or both negative (when $a < b$). Hence we always have

$$(a^m - b^m)(a^n - b^n) \geq 0$$

with equality if and only if $a = b$. Rearranging this inequality as

$$a^{m+n} + b^{m+n} \geq a^n b^m + a^m b^n ,$$

adding $a^{m+n} + b^{m+n}$ to both sides, and factoring the right-hand side, we get

$$2(a^{m+n} + b^{m+n}) \geq (a^m + b^m)(a^n + b^n)$$

or

$$\frac{a^{m+n} + b^{m+n}}{2} \geq \frac{a^m + b^m}{2} \cdot \frac{a^n + b^n}{2} .$$

Equality holds if and only if $a = b$.   □

Ordering can help us establish an inequality that is symmetric in its variables. Consider, for instance, the inequality [36]

$$(a^2 + b^2 + c^2)(a^3 + b^3 + c^3) \leq 3(a^5 + b^5 + c^5) \qquad (1.41)$$

for positive real variables a, b, c. Because of the symmetric way in which these appear, there is no loss of generality in assuming that $a \leq b \leq c$. This will eventually permit us to recognize (1.41) as a special case of Chebyshev's inequality for sums (Theorem 3.8).

1.8 Problems

1.1. Prove Theorem 1.1.

1.2. Let $w, x, y, z \in \mathbb{R}$. Use the fundamental inequality (1.30) to prove

(a) $(x + y)^2 \geq 4xy$ with equality if and only if $x = y$,

(b) $(wy - xz)^2 \geq (w^2 - x^2)(y^2 - z^2)$,

(c) $x^2 y^2 + y^2 z^2 + z^2 x^2 \geq xyz(x + y + z)$,

(d) $x(x - y) \geq y(x - y)$,

(e) $x^4 + y^4 + z^4 \geq xyz(x + y + z)$,

(f) $2xyz \leq x^2 + y^2 z^2$,

(g) $(x + y + z)^2 \geq 3(xy + yz + zx)$.

1.3. Show that

$$x + \frac{1}{x} \geq 2 \qquad (x > 0)$$

with equality if and only if $x = 1$.

1.4. Prove that

$$|ab| \leq \frac{\varepsilon a^2}{2} + \frac{b^2}{2\varepsilon} \qquad (\varepsilon > 0) .$$

1.5. [68, 73] Let $f(x)$ and $g(x)$ be real functions of a real variable x. Outline a general strategy for solving an inequality of the form

(a) $\sqrt[2n+1]{f(x)} > g(x)$,

(b) $\sqrt[2n+1]{f(x)} < g(x)$,

(c) $\sqrt[2n]{f(x)} > g(x)$,

(d) $\sqrt[2n]{f(x)} < g(x)$,

(e) $\sqrt[2n]{f(x)}/g(x) > 1$,

(f) $\sqrt[2n]{f(x)}/g(x) < 1$.

1.6. [68, 73] Let $f(x)$, $g(x)$, and $h(x)$ be real functions. Discuss the solution sets of the following inequality types. Take a, b, α to be real constants.

(a) $a^x > b$ and $a^x < b$, where $a > 0$.

(b) $a^x > a^\alpha$ where $a > 0$.

(c) $a^{f(x)} > a^{g(x)}$ where $a > 0$.

(d) $f(x)^{h(x)} < f(x)^{g(x)}$.

(e) $f(x)^{h(x)} < g(x)^{h(x)}$.

(f) $\log_a x > b$ and $\log_a x < b$, where $a > 0$.

(g) $\log_a x > \log_a \alpha$ and $\log_a x < \log_a \alpha$, where $a > 0$ and $\alpha > 0$.

(h) $\log_a f(x) < \log_a g(x)$ where $a > 0$.

(i) $\log_{g(x)} f(x) > 0$ and $\log_{g(x)} f(x) < 0$.

(j) $\log_{h(x)} f(x) < \log_{h(x)} g(x)$.

1.7. Solve each inequality, treating a as a fixed but arbitrary real parameter.

(a) $x^2 - 2x - 3 < 0$,

(b) $(x^2 - 16)(x + 4)(x - 1) > 0$,

(c) $(x - 1)/(x + 1) < 0$,

(d) $x + 1/x < 0$,

(e) $(x - 1)/(x + 1) < x$,

(f) $x^2 - |x| - 1 \geq 0$,

(g) $|x^2 - x| < x$,

(h) $|x^3 - 1| > 1 - x$,

(i) $1/|x + 1| < 1/|x - 1|$,

(j) $\sqrt{x + 2} > x$,

(k) $\sqrt{x^2} < x + 1$,

(l) $x < \sqrt{1 - |x|}$,

(m) $(1/2)^{x+1} \geq 1$,

(n) $(\log_2 x^2)^2 \leq 4$,

(o) $(x + 1)^x < (x + 1)^{2-x}$,

(p) $|x| < a/x$,

(q) $ax > 1$,

(r) $ax > 1/x$,

(s) $ax^2 + 1 \geq 0$,

(t) $|x| \geq x - a$,

(u) $|x^2 - 1| > a$,

(v) $a\sqrt{x + 1} < 1$,

(w) $|x - a| \geq x + 1$,

(x) $|x| + 1 > |ax|$,

(y) $|x^2 - 1| < a$,

(z) $\log_2 x < a + 2$.

1.8. Use mathematical induction to prove the following [43, 81, 83]:

(a) $2^{n+2} > 2n + 5$ for $n \geq 1$,

(b) $2^n > 2n + 1$ for $n \geq 3$,

(c) $\sum_{k=1}^{n} 1/\sqrt{k} < 2\sqrt{n}$,

(d) $(1/2) \cdot (3/4) \cdot (5/6) \cdots [(2n-1)/(2n)] \leq 1/\sqrt{3n+1}$,

(e) $2! \, 4! \cdots (2n)! > [(n+1)!]^n$ for $n \geq 2$,

(f) $4^n(n!)^2 < (n+1)(2n)!$ for $n \geq 2$,

(g) $a_2^2/a_1 + a_3^2/a_2 + \cdots + a_n^2/a_{n-1} \geq 4(a_n - a_1)$ for $a_1, \ldots, a_n > 0$ and $n \geq 2$,

(h) $\sum_{k=1}^{n} 1/(k\sqrt{k}) \leq 3 - 2/\sqrt{n}$ for $n \geq 1$,

(i) $1/2^2 + 1/3^2 + \cdots + 1/n^2 < 1$ for $n \geq 2$,

(j) $2^{\frac{1}{2}k(k-1)} > k!$ for $k \geq 3$,

(k) $(1 + x)^n \leq 1 + (2^n - 1)x$ for $n \in \mathbb{N}$ and $0 \leq x \leq 1$,

(l) $n! > 2^{n-1}$ for $n > 2$,

(m) $2^{n+1} > n^2$.

1.9. Let a_i/b_i ($i = 1, \ldots, n$) be fractions having denominators b_i all positive. Show that

$$\min_{1 \leq i \leq n} \frac{a_i}{b_i} \leq \frac{\sum_{i=1}^{n} a_i}{\sum_{i=1}^{n} b_i} \leq \max_{1 \leq i \leq n} \frac{a_i}{b_i} \, .$$

1.10. Establish the following two bounds for differences of powers [36, 54, 83]:

(a) $n(x-y)y^{n-1} < x^n - y^n < n(x-y)x^{n-1}$ for $0 \leq y < x$ and $n \geq 2$,

(b) $|x^n - y^n| \leq n|x-y| \, (\max\{|x|, |y|\})^{n-1}$.

1.11. Prove the following statements (known as *Weierstrass's inequalities*).

(a)   For positive real numbers $a_1, \ldots, a_n$, we have

$$\prod_{i=1}^{n}(1 + a_i) \geq 1 + \sum_{i=1}^{n} a_i \, .$$

(b)   For $n \geq 2$ with $0 \leq a_i < 1$, we have

$$\prod_{i=1}^{n}(1 - a_i) > 1 - \sum_{i=1}^{n} a_i \, .$$

For related inequalities and applications to convergence of infinite products, see Bromwich [12].

1.12. The *Fibonacci numbers* f_n are defined by the recursion $f_n = f_{n-1} + f_{n-2}$ with $f_1 = f_2 = 1$. Show that $f_n < 2^n$ for $n \in \mathbb{N}$.

1.13. Show that with $z = x + iy$, we have $|\sinh y| \leq |\cos z| \leq \cosh y$.

1.14. Given positive numbers $a_1, \ldots, a_m$, the numbers A, H, and G defined by

$$A = \frac{1}{m} \sum_{n=1}^{m} a_n, \qquad H = \left(\frac{1}{m} \sum_{n=1}^{m} a_n^{-1} \right)^{-1}, \qquad G = \left(\prod_{n=1}^{m} a_n \right)^{1/m},$$

are the *arithmetic*, *harmonic*, and *geometric* means, respectively, of the set. Show that if the a_n are not all equal, then each of the means lies between the minimum and maximum values of the a_n.

1.15. Prove the following assertions about suprema and infima. Assume all sets are subsets of $\mathbb{R}$.

(a) We have $s = \sup S$ if and only if: (1) $x \leq s$ for all $x \in S$; and (2) for all $\varepsilon > 0$, there exists $y \in S$ such that $y > s - \varepsilon$.

(b) Let $A \subseteq B$. If $\sup A$ and $\sup B$ exist, then $\sup A \leq \sup B$. Similarly, $\inf A \geq \inf B$.

(c) If $x < M$ for all $x \in S$, then $\sup S \leq M$. Similarly, if $x > m$ for all $x \in S$, then $\inf S \geq m$. Note that the process of taking sup or inf can *blunt* an inequality (change it from strict to weak).

1.16. Assume that $f(x)$ and $g(x)$ are defined over a common domain D, and establish the following relations. (Take all subsets to be nonempty.)

(a) If $S_1 \subseteq S_2 \subseteq D$, then $\sup_{x \in S_1} f(x) \leq \sup_{x \in S_2} f(x)$ and $\inf_{x \in S_1} f(x) \geq \inf_{x \in S_2} f(x)$.

(b) Assume $f(x) \geq g(x)$ for all $x \in S \subseteq D$.

   (i) If g is bounded below on S, then $\inf_{x \in S} f(x) \geq \inf_{x \in S} g(x)$.

   (ii) If f is bounded above on S, then $\sup_{x \in S} f(x) \geq \sup_{x \in S} g(x)$.

(c) If $S \subseteq D$, then $\sup_{x \in S}[f(x) + g(x)] \leq \sup_{x \in S} f(x) + \sup_{x \in S} g(x)$.

1.17. Use homogeneity to prove the following for $a, b > 0$:

(a) $1/a + 1/b \leq a/b^2 + b/a^2$,

(b) $\sqrt{a^2/b} + \sqrt{b^2/a} \geq \sqrt{a} + \sqrt{b}$,

(c) $a^5 + b^5 \geq a^2 b^3 + a^3 b^2$.

1.18. Use symmetry in the variables to prove the inequality

$$|a + b|^p \leq 2^p (|a|^p + |b|^p) \qquad (p \geq 1).$$

1.19. (*Rough bound for zeros of a polynomial*). Suppose

$$f(z) = a_0 z^n + a_1 z^{n-1} + \cdots + a_{n-1} z + a_n \qquad (a_0 \neq 0)$$

where z is complex and the coefficients a_i may be real or complex. Show that all the zeros of $f(z)$ have moduli less than or equal to the number

$$\xi = 1 + \frac{n}{|a_0|} \cdot \max_{1 \leq k \leq n} |a_k|. \tag{1.42}$$

Chapter 2
Methods from the Calculus

2.1 Introduction

In this chapter we revisit some facts from mathematical analysis and show how these may be used to establish important inequalities. We begin by reviewing convergence of real number sequences and continuity of real functions of a single variable.

2.2 Limits and Continuity

Convergent Sequences of Real Numbers

We say that a real sequence $\{x_n\}$ is bounded above if there exists M such that $x_n \le M$ for all n. It is bounded below if there exists m such that $x_n \ge m$ for all n, and it is *bounded* if it is bounded above and bounded below (i.e., if there exists B such that $|x_n| \le B$ for all n). Although unbounded sequences can be fascinating, our main interest will be in bounded sequences.

We say that $\{x_n\}$ is a Cauchy sequence if for every positive number ε there exists a positive integer N (dependent on ε) such that

$$|x_n - x_m| < \varepsilon \ \text{ whenever } \ m, n > N .$$

A sequence $\{x_n\}$ is convergent and has limit x if for every $\varepsilon > 0$ there exists $N \in \mathbb{N}$ such that

$$|x_n - x| < \varepsilon \tag{2.1}$$

whenever $n > N$. In this case we write

$$\lim_{n \to \infty} x_n = x \quad \text{or} \quad x_n \to x \text{ as } n \to \infty .$$

The limit of a convergent sequence is unique.

M.J. Cloud et al., *Inequalities: With Applications to Engineering*,
DOI 10.1007/978-3-319-05311-0_2, © Springer International Publishing AG 2014

Remark 2.1. We have $x_n \to x$ if and only if for each $\varepsilon > 0$, the solution set with respect to n of the inequality (2.1) contains an interval of the form $(N(\varepsilon), \infty)$. □

We say that $\{x_n\}$ is increasing if $x_{n+1} \geq x_n$ for all n, or decreasing if $x_{n+1} \leq x_n$ for all n. If strict inequality holds we use the terms strictly increasing or decreasing, respectively. An increasing sequence of real numbers is convergent if and only if it is bounded above, and a decreasing sequence is convergent if and only if it is bounded below. Let $\{x_n\}$ be a bounded sequence. Define

$$a_n = \inf_{k \geq n} x_k , \qquad b_n = \sup_{k \geq n} x_k .$$

Then $\{a_n\}$ is increasing and bounded above, while $\{b_n\}$ is decreasing and bounded below. The two numbers defined respectively by

$$\underline{\lim}_{n \to \infty} x_n = \lim_{n \to \infty} a_n , \qquad \overline{\lim}_{n \to \infty} x_n = \lim_{n \to \infty} b_n ,$$

are the *limit inferior* and the *limit superior* of $\{x_n\}$. Although we have

$$a_1 \leq a_2 \leq a_3 \leq \cdots \leq b_3 \leq b_2 \leq b_1 ,$$

we are not guaranteed that $\{x_n\}$ is convergent. The sequence $\{(-1)^k\}$, for example, oscillates between -1 (its limit inferior) and $+1$ (its limit superior). However, if the limit inferior and the limit superior of a bounded sequence happen to coincide as a number x, then the sequence has limit x.

Sequences receive extensive coverage in any standard calculus text. There are many useful results in the subject (e.g., the various tests for convergence and divergence) and a number of these serve as interesting applications of inequalities (e.g., the comparison tests). We will assume a working knowledge of the basic theorems on sequence limits (the limit of a sum is the sum of the limits, etc.). The following two results, however, are central to our purposes.

Lemma 2.1 (Limit Passage). *If $\{x_n\}$ and $\{y_n\}$ are real sequences such that $x_n \to x$ and $y_n \to y$ as $n \to \infty$ with $x_n \leq y_n$ for all n, then $x \leq y$.*

Proof. Let $\varepsilon > 0$ be given and choose N_1 and N_2 so that $n \geq \max(N_1, N_2)$ implies $x - \varepsilon/2 < x_n$ and $y_n < y + \varepsilon/2$. The inequality $x - \varepsilon/2 < x_n \leq y_n < y + \varepsilon/2$ shows that $x - y < \varepsilon$, and since $\varepsilon > 0$ is arbitrary we have $x - y \leq 0$. Hence $x \leq y$. □

Note that, in general, an inequality may be blunted by a limit passage. That is, we may have $x_n < y_n$ for all n but $x \leq y$. Consider $x_n = 0$ and $y_n = 1/n$, for example.

Lemma 2.2 (Squeeze Principle). *If $a_n \to L$ and $c_n \to L$ as $n \to \infty$ and there exists N such that $a_n \leq b_n \leq c_n$ for all $n > N$, then $b_n \to L$ as $n \to \infty$.*

Proof. Let $\varepsilon > 0$ be given. There exists M such that $n > M$ implies

$$L - \varepsilon < a_n \leq b_n \leq c_n < L + \varepsilon .$$

Hence $|b_n - L| \leq \varepsilon$ for all $n > M$. □

Example 2.1. In Russia, the squeeze principle is commonly called the *policemen theorem*: $\{a_n\}$ and $\{c_n\}$ are described as "policeman" sequences who funnel "criminal" sequence $\{b_n\}$ toward a police station. Let us combine the squeeze principle with a nonreversible transformation. We can discard almost all the terms from the binomial expansion

$$(1+x)^n = 1 + nx + \frac{n(n-1)}{2!}x^2 + \cdots \tag{2.2}$$

and write, for instance,

$$(1+x)^n > \frac{n(n-1)}{2}x^2 \qquad (x > 0,\ n \in \mathbb{N},\ n > 1). \tag{2.3}$$

To show that

$$\lim_{n\to\infty} \frac{n}{a^n} = 0 \qquad (a > 1), \tag{2.4}$$

we can use (2.3) to write

$$\frac{n}{a^n} = \frac{n}{[1+(a-1)]^n} < \frac{n}{\dfrac{n(n-1)}{2}(a-1)^2} = \frac{2}{(n-1)(a-1)^2}.$$

Then

$$0 < \frac{n}{a^n} < \frac{2}{(n-1)(a-1)^2} \qquad (n > 1)$$

and Lemma 2.2 gives (2.4). $\qquad\qquad\qquad\qquad\qquad\qquad\qquad\qquad\qquad\qquad\square$

Limits and Continuity for Real Functions of a Single Variable

Let $f = f(x)$ be a real-valued function of the real variable x. We say that f has limit L as $x \to x_0$ if for every $\varepsilon > 0$ there exists $\delta > 0$ (dependent on ε) such that $|f(x) - L| < \varepsilon$ whenever $|x - x_0| < \delta$. We assume a working knowledge of the basic limit theorems for functions (the limit of a product is the product of the limits, and so on). Lemmas 2.1 and 2.2 have their counterparts for functions. For example, if $g(x) \to L$ and $h(x) \to L$ as $x \to x_0$, with $g(x) \le f(x) \le h(x)$, then $f(x) \to L$ as $x \to x_0$. To be complete, however, we would have to state several additional cases for functions. For instance, f has limit L as $x \to +\infty$ if for every $\varepsilon > 0$ there exists N such that $|f(x) - L| < \varepsilon$ whenever $x > N$. The squeeze principle could be rephrased accordingly.

The statement that f is *continuous* at $x = x_0$ means that for every $\varepsilon > 0$, there is a $\delta > 0$ such that $|f(x) - f(x_0)| < \varepsilon$ whenever $|x - x_0| < \delta$. We say that f is continuous on an interval I if f is continuous at every $x \in I$ (with suitable modifications made for continuity at endpoints of closed intervals). Two useful facts about continuity are the following.

Lemma 2.3 (Persistence of Sign). *Suppose f is continuous at $x = x_0$ and $f(x_0)$ is nonzero. Then there is an open interval containing x_0 such that $f(x)$ is nonzero at every point of the interval.*

Proof. Assume $f(x_0) > 0$ (otherwise replace f by $-f$). Let $\varepsilon = f(x_0)$. There exists $\delta > 0$ such that $|x - x_0| < \delta$ implies $|f(x) - f(x_0)| < \varepsilon$, so if $x \in (x_0 - \delta, x_0 + \delta)$ then $-\varepsilon < f(x) - f(x_0) < \varepsilon$. Hence $f(x_0) - \varepsilon < f(x) < f(x_0) + \varepsilon$ or, since $f(x_0) = \varepsilon$, we have $0 < f(x)$. $\qquad\qquad\qquad\qquad\qquad\qquad\qquad\qquad\qquad\qquad\qquad\qquad\qquad\square$

Theorem 2.1 (Sequential Continuity). *A function f is continuous at x_0 if and only if $f(x_n) \to f(x_0)$ whenever $x_n \to x_0$.*

Proof. Suppose f is continuous at x_0 and $x_n \to x_0$. Let $\varepsilon > 0$. There exists $\delta > 0$ such that $|x - x_0| < \delta$ implies $|f(x) - f(x_0)| < \varepsilon$. Now suppose $x_n \to x_0$. Choose N such that $n > N$ implies $|x_n - x_0| < \delta$. For this N, $n > N$ implies $|f(x_n) - f(x_0)| < \varepsilon$ and therefore $f(x_n) \to f(x_0)$. Conversely, suppose $f(x_n) \to f(x_0)$ whenever $x_n \to x_0$. To show that f is continuous at x_0, we suppose f is not continuous at x_0 and seek a contradiction. There exists $\varepsilon > 0$ such that for any $\delta > 0$, there exists some x with $|x - x_0| < \delta$ but $|f(x) - f(x_0)| \geq \varepsilon$. In particular we may choose a sequence $\delta_i = 1/i$ and x_i with $|x_i - x_0| < \delta_i$ but $|f(x_i) - f(x_0)| \geq \varepsilon$ for all $i \in \mathbb{N}$. Then $x_i \to x_0$, but it is false that $f(x_i) \to f(x_0)$. $\qquad\qquad\qquad\qquad\qquad\qquad\qquad\qquad\qquad\qquad\square$

Theorem 2.1, sometimes called *Heine's theorem*, provides a notion of continuity equivalent to the less intuitive ε-δ definition. However, the ε-δ definition can be more convenient in proving continuity of a particular function, as it reduces to solving the inequality $|f(x) - f(x_0)| < \varepsilon$ and demonstrating that the solution contains the interval $|x - x_0| < \delta$ for some small δ. Note that to prove that f is *not* continuous at x_0, it suffices to exhibit a sequence $x_k \to x_0$ such that $f(x_k) \not\to f(x_0)$ as $k \to \infty$.

The consequences of continuity on a closed interval are particularly important. We state the following without proof, referring the reader to any standard calculus text for details. One of these consequences is known as the *intermediate value property*.

Theorem 2.2. *If f is continuous on $[a, b]$, then $f(x)$ assumes every value between $f(a)$ and $f(b)$, $f(x)$ is bounded on $[a, b]$, and $f(x)$ takes on its supremum and its infimum in $[a, b]$.*

Finally, we review concepts relating to monotonicity and extrema. A function f is increasing on I if $f(x_2) \geq f(x_1)$ whenever $x_2 > x_1$ for all $x_1, x_2 \in I$. Similarly, f is decreasing if $f(x_2) \leq f(x_1)$ whenever $x_2 > x_1$. If strict inequality holds we use the terms strictly increasing or decreasing, respectively. Let $x_0 \in I$. If $f(x_0) \geq f(x)$ for all $x \in I$, then f has a maximum on I equal to $f(x_0)$. The definition of minimum is analogous.

2.3 Basic Results for Integrals

The formal definition of the Riemann integral appears in Problem 2.2. It is helpful to keep in mind that $f = f(x)$ is integrable on $[a, b]$ if it is continuous or monotonic on $[a, b]$.

Several useful inequalities for integrals can be established by forming Riemann sums. Given an integral

$$\int_a^b f(x)\, dx\,,$$

we use the notation $\Delta x = (b - a)/n$ and $x_i = a + i\Delta x$ for $i = 0, \ldots, n$, and write the corresponding Riemann sum as

$$\sum_{i=1}^n f(x_i)\, \Delta x\,.$$

Once an inequality is established for such a sum, we may let $n \to \infty$ and, being assured of convergence of the Riemann sums to the integral, apply Lemma 2.1 in order to obtain an inequality involving the integral.

Theorem 2.3. *If f and g are integrable on $[a, b]$ with $f(x) \le g(x)$, then*

$$\int_a^b f(x)\, dx \le \int_a^b g(x)\, dx\,.$$

Proof. With the notation described above, we form Riemann sums for the integrals. It is seen that

$$\sum_{i=1}^n f(x_i)\, \Delta x \le \sum_{i=1}^n g(x_i)\, \Delta x\,.$$

The result follows as $n \to \infty$ by Lemma 2.1.   □

Example 2.2. A simple observation shows that

$$\int_t^\infty \frac{e^{-x^2}}{x^{2n}}\, dx = \int_t^\infty \frac{xe^{-x^2}}{x^{2n+1}}\, dx \le \int_t^\infty \frac{xe^{-x^2}}{t^{2n+1}}\, dx = \frac{e^{-t^2}}{2t^{2n+1}}\,.$$

We used the fact that $1/x^{2n+1} \le 1/t^{2n+1}$ for $x \in [t, \infty)$.   □

Corollary 2.1 (Simple Estimate). *If f is integrable on $[a, b]$ with $m \le f(x) \le M$, then*

$$m(b - a) \le \int_a^b f(x)\, dx \le M(b - a)\,. \tag{2.5}$$

Consequently, the average value of f over $[a, b]$ lies between m and M.

Example 2.3. We have

$$(b-a)\sqrt{ca^3+d} < \int_a^b \sqrt{cx^3+d}\,dx < (b-a)\sqrt{cb^3+d}$$

for any positive constants a, b, c, d with $a < b$.   □

Corollary 2.2 (Modulus Inequality). *If f is integrable on $[a, b]$, then*

$$\left|\int_a^b f(x)\,dx\right| \le \int_a^b |f(x)|\,dx . \tag{2.6}$$

This follows from the inequalities $-|f(x)| \le f(x) \le |f(x)|$ and plays the role of the triangle inequality for integrals.

If continuity is assumed in the integrand function f, the persistence of sign property leads to the next result.

Lemma 2.4. *Let f be continuous on $[a, b]$ and suppose $f(x) \ge 0$ on $[a, b]$ with $f(x_0) > 0$ for some $x_0 \in [a, b]$. Then*

$$\int_a^b f(x)\,dx > 0 .$$

Proof. If $x_0 \in (a, b)$, there is an open interval about x_0 where $f(x) > 0$. Choose a smaller closed interval where $f(x) > 0$, say $I = [x_0 - \delta, x_0 + \delta]$. Let m be the minimum value of $f(x)$ in I. Then

$$\int_a^b f(x)\,dx \ge m(2\delta) > 0 .$$

If x_0 is an endpoint, then $f(x)$ is also positive at an interior point so the argument still applies.   □

A class of results known as mean value theorems are also useful. We give two of these and refer the reader to Problem 2.5 for other examples.

Theorem 2.4 (Second Mean Value Theorem for Integrals). *If f is continuous on $[a, b]$, and g is integrable and never changes sign on $[a, b]$, then for some $\xi \in [a, b]$*

$$\int_a^b f(x)g(x)\,dx = f(\xi) \int_a^b g(x)\,dx . \tag{2.7}$$

Proof. Assume that $g(x) \ge 0$ on $[a, b]$; otherwise, replace $g(x)$ by $-g(x)$. Let M and m denote the maximum and minimum values, respectively, of $f(x)$ on $[a, b]$. Then

$$mg(x) \le f(x)g(x) \le Mg(x)$$

for all x, hence

$$m \int_a^b g(x)\,dx \le \int_a^b f(x)g(x)\,dx \le M \int_a^b g(x)\,dx \,.$$

If $\int_a^b g(x)\,dx = 0$ then any choice of ξ will do. Otherwise

$$m \le \frac{\int_a^b f(x)g(x)\,dx}{\int_a^b g(x)\,dx} \le M \,.$$

By the intermediate value property applied to f,

$$f(\xi) = \frac{\int_a^b f(x)g(x)\,dx}{\int_a^b g(x)\,dx}$$

for some $\xi \in [a, b]$, and (2.7) follows.

Corollary 2.3 (First Mean Value Theorem for Integrals). *If f is continuous on $[a, b]$, then for some $\xi \in [a, b]$*

$$\int_a^b f(x)\,dx = f(\xi)(b - a) \,.$$

Hence $f(\xi)$ equals the average value of $f(x)$ on $[a, b]$.

Example 2.4. Consider the integral

$$I = \int_0^1 \frac{x^5}{(x + 25)^{1/2}}\,dx \,.$$

On the interval $[0, 1]$ we have $x^5 \ge 0$; hence by Theorem 2.4 there exists $\xi \in [0, 1]$ such that

$$I = \frac{1}{6(\xi + 25)^{1/2}} \,.$$

Therefore $1/(6\sqrt{26}) \le I \le 1/30$. $\square$

Through a process reminiscent of the integral test for series, we can obtain other inequalities involving integrals.

Example 2.5. The function $f(x) = x^p$ ($-1 < p < 0$) is strictly decreasing on $(0, \infty)$. From Fig. 2.1 it is apparent that

$$\int_1^{n+1} x^p\,dx < \sum_{k=1}^n k^p < \int_0^n x^p\,dx \,.$$

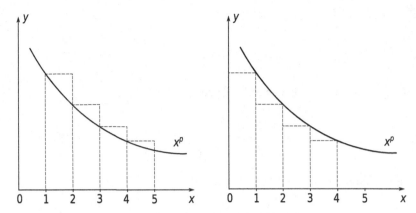

Fig. 2.1 Comparing a sum to two integrals (Example 2.5 with $n = 4$)

Hence, after carrying out the integrations, we get

$$\frac{(n + 1)^{p+1} - 1}{p + 1} < \sum_{k=1}^{n} k^p < \frac{n^{p+1}}{p + 1} .$$

□

Integration along a contour in the complex plane follows many rules analogous to those for real integration, with little modification. In particular, Corollary 2.2 extends to the complex case: if $g(z)$ is integrable on contour C, then

$$\left| \int_C g(z) \, dz \right| \leq \int_C |g(z)| \, |dz| .$$

Example 2.6. Suppose C is of finite length L. If there is a number $M > 0$ such that $|g(z)| < M$ for all $z \in C$, then

$$\left| \int_C g(z) \, dz \right| \leq \int_C |g(z)| \, |dz| < \int_C M \, |dz| = M \int_C |dz| = ML . \qquad (2.8)$$

This is sometimes called the *Darboux inequality*.   □

2.4 Results from the Differential Calculus

A function f is said to be n times continuously differentiable on an interval I if the first n derivatives of f exist and are continuous on I. We first recall the Fundamental Theorem of Calculus. A proof may be found in any standard calculus text.

Theorem 2.5. *If f is continuous on $[a, b]$ and $F'(x) = f(x)$, then*

$$\int_a^b f(x) \, dx = F(b) - F(a) .$$

The next result is a source of series expansions that are useful in approximating functions and, as we shall see, in generating inequalities.

Theorem 2.6 (Taylor's Theorem). *Let $x > a$, suppose f is n times continuously differentiable on $[a, x]$, and suppose $f^{(n+1)}(x)$ exists on (a, x). Then*

$$f(x) = f(a) + f'(a)(x - a) + \cdots + \frac{f^{(n)}(a)}{n!}(x - a)^n + \frac{f^{(n+1)}(\xi)}{(n + 1)!}(x - a)^{n+1}$$

for some ξ strictly between a and x.

Proof. The first $n + 1$ terms constitute the Taylor polynomial of degree n for $f(x)$ about the point a; the last term is called the remainder term. To simplify the proof, assume f is $n + 1$ times continuously differentiable on $[a, b]$. By Theorem 2.5,

$$f(x) - f(a) = \int_a^x f'(t)\, dt \ .$$

Integrate by parts with $u = f'(t)$, $du = f''(t)\, dt$, $v = -(x - t)$, and $dv = dt$; then

$$\int_a^x f'(t)\, dt = f'(a)(x - a) + \int_a^x f''(t)(x - t)\, dt \ .$$

Repeat with $u = f''(t)$, $du = f'''(t)\, dt$, $v = -(x - t)^2/2$, $dv = (x - t)\, dt$, and continue the process until

$$f(x) = f(a) + f'(a)(x - a) + \cdots + \frac{f^{(n)}(a)}{n!}(x - a)^n + \frac{1}{n!}\int_a^x f^{(n+1)}(t)(x - t)^n\, dt \ .$$

Because $(x - t)^n$ never changes sign in the interval with endpoints a and x, by (2.7) the remainder term can be rewritten

$$\frac{1}{n!}\int_a^x f^{(n+1)}(t)(x - t)^n\, dt = \frac{f^{(n+1)}(\xi)}{n!}\int_a^x (x - t)^n\, dt = \frac{f^{(n+1)}(\xi)}{(n + 1)!}(x - a)^{n+1}$$

for some ξ between a and x.   □

We can sometimes establish inequalities through inspection of series expansions.

Example 2.7. From the Taylor series

$$e^x = \sum_{n=0}^{\infty} \frac{x^n}{n!}$$

we see that

$$e^x > 1 + x + \tfrac{1}{2}x^2 \qquad (x > 0) \ .$$

Even more simply we have $e^x > 1 + x$, but we can replace x by x/n to get the less obvious result

$$e^x > \left(1 + \frac{x}{n}\right)^n \qquad (x > 0, \ n \in \mathbb{N}) .$$

If $z \in \mathbb{C}$, relation (1.17) yields

$$|\sin z| = \left|\sum_{n=1}^{\infty} (-1)^{n-1} \frac{z^{2n-1}}{(2n-1)!}\right| \leq \sum_{n=1}^{\infty} \left|(-1)^{n-1} \frac{z^{2n-1}}{(2n-1)!}\right| = \sum_{n=1}^{\infty} \frac{|z|^{2n-1}}{(2n-1)!}$$

and we have $|\sin z| \leq \sinh |z|$.   □

The next two results, although important in their own right, can be viewed as immediate consequences of Taylor's theorem.

Corollary 2.4 (Mean Value Theorem). *Suppose f is continuous on $[a, b]$ and differentiable in (a, b). Then there exists $\xi \in (a, b)$ such that*

$$f(b) = f(a) + f'(\xi)(b - a) . \tag{2.9}$$

Intuitively, there is a point in (a, b) such that the slope of the line tangent to $f(x)$ at that point equals the slope of the secant line connecting the function values at the endpoints of $[a, b]$. See Fig. 2.2.

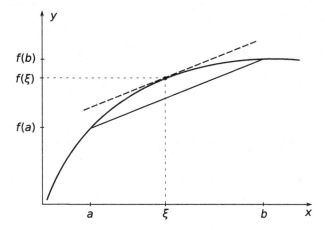

Fig. 2.2 Mean value theorem. The heavier *dashed line* is tangent to $y = f(x)$ at $x = \xi$

Example 2.8. We can verify the inequality

$$\tan x > x \qquad (0 < x < \pi/2) \tag{2.10}$$

by applying Corollary 2.4 with $f(x) = \tan x$, $a = 0$, and $b = x < \pi/2$; i.e., by asserting that

$$\tan x - \tan 0 = \frac{1}{\cos^2 \xi}(x - 0) \ \text{ for some } \ \xi \in (0, x)$$

and noting that $0 < \cos^2 \xi < 1$. Similarly, we have $\sin x < x$ for $x > 0$ so that

$$\sin x < x < \tan x \qquad (0 < x < \pi/2) \,.$$

Applying Corollary 2.4 to the natural log, we obtain

$$\ln(1 + x) - \ln 1 = \frac{1}{\xi}[(1 + x) - 1] \;\text{ for some }\; \xi \in (1, 1 + x) \,.$$

Therefore

$$\frac{x}{1 + x} < \ln(1 + x) < x \qquad (x > 0) \,. \qquad\qquad (2.11)$$

This is the *logarithmic inequality*. It also holds for $-1 < x < 0$. $\qquad\qquad \square$

Corollary 2.5 (Rolle's Theorem). *If f is continuous on $[a, b]$ and differentiable in (a, b) with $f(b) = f(a) = 0$, then there exists $\xi \in (a, b)$ such that $f'(\xi) = 0$.*

Rolle's theorem indicates that between every two zeros of a continuous function the derivative has at least one zero.

Theorem 2.7 (Conditions for Monotonicity). *If f is continuous on $[a, b]$ and differentiable in (a, b) with $f'(x) \geq 0$, then f is increasing on $[a, b]$. If $f'(x) > 0$, then f is strictly increasing. Corresponding statements hold for decreasing functions, for which $f'(x) \leq 0$.*

Proof. We prove only the first part of the theorem, and leave the rest for the reader. Suppose $a \leq x_1 < x_2 \leq b$. By Corollary 2.4, there is a number $\xi \in (x_1, x_2)$ such that $f(x_2) - f(x_1) = f'(\xi)(x_2 - x_1)$. But $f'(\xi) \geq 0$ by hypothesis and $x_2 - x_1 > 0$, so $f(x_2) - f(x_1) \geq 0$. Hence $f(x_2) \geq f(x_1)$ whenever $x_2 > x_1$ on $[a, b]$. $\qquad \square$

Example 2.9. The average of a positive, increasing function is increasing. For let $f(x)$ be increasing on $[0, a]$. Then for every $x \in (0, a]$ we have

$$f(x) \geq \max_{u \in [0,x]} f(u) = \max_{u \in [0,x]} f(u) \cdot \frac{1}{x} \int_0^x du \geq \frac{1}{x} \int_0^x f(u)\, du \,.$$

Hence

$$f(x) - \frac{1}{x} \int_0^x f(u)\, du \geq 0 \quad \text{so that} \quad \frac{1}{x^2}\left(x f(x) - \int_0^x f(u)\, du\right) \geq 0 \,.$$

By the quotient rule for differentiation,

$$\frac{d}{dx}\left(\frac{1}{x} \int_0^x f(u)\, du\right) \geq 0$$

as required. $\qquad\qquad \square$

Theorem 2.8 (Cauchy's Mean Value Theorem). *Suppose f, g are continuous on $[a, b]$ and differentiable in (a, b). Then there exists $\xi \in (a, b)$ such that*

$$[f(b) - f(a)] g'(\xi) = [g(b) - g(a)] f'(\xi) .$$

Proof. Call $A = f(b) - f(a)$, $B = -[g(b) - g(a)]$, $C = -Bf(a) - Ag(a)$, and apply Rolle's theorem to $h(x) = Ag(x) + Bf(x) + C$. □

The following is useful for establishing the monotonicity of the ratio of two functions.

Theorem 2.9 (l'Hôpital's Monotone Rule). *Suppose f, g are continuous on $[a, b]$ and differentiable in (a, b) with $g'(x) \neq 0$ in (a, b). Let $f'(x)/g'(x)$ be increasing (or decreasing) on (a, b). Then the functions*

$$\frac{f(x) - f(a)}{g(x) - g(a)} \quad and \quad \frac{f(x) - f(b)}{g(x) - g(b)} \tag{2.12}$$

are also increasing (or decreasing) on (a, b). If $f'(x)/g'(x)$ is strictly increasing (or decreasing) so are the functions (2.12).

Proof. (See [3]). We may assume $g'(x) > 0$ on (a, b). (If not, multiply f and g by -1.) By Theorem 2.8, for $x \in (a, b)$ there exists $y \in (a, x)$ with

$$\frac{f(x) - f(a)}{g(x) - g(a)} = \frac{f'(y)}{g'(y)} \leq \frac{f'(x)}{g'(x)} , \quad so \quad f'(x) \geq g'(x) \frac{f(x) - f(a)}{g(x) - g(a)} .$$

Now use the quotient rule and the last expression to deduce that the derivative of $[f(x) - f(a)]/[g(x) - g(a)]$ is nonnegative, hence Theorem 2.7 applies. □

By l'Hôpital's rule, to evaluate a ratio of the indeterminate form $0/0$ we differentiate both numerator and denominator and try to evaluate again. Theorem 2.9 is almost as easily remembered. To establish that a ratio is monotone on an interval using Theorem 2.9, we verify that we get $0/0$ at one of the endpoints, then differentiate numerator and denominator and check that the resulting quotient is monotone (making sure the new denominator is nonzero on the open interval).

Theorem 2.10 (Second Derivative Test). *Assume f is twice continuously differentiable in (a, b). Let $x_0 \in (a, b)$, and suppose that $f'(x_0) = 0$ and $f''(x_0) > 0$. Then f has a local minimum at x_0. That is, there exists $\delta > 0$ such that $0 < |x - x_0| < \delta$ implies $f(x) > f(x_0)$.*

Proof. Because $f''(x_0) > 0$, by Lemma 2.3 there exists $\delta > 0$ such that $f''(x) > 0$ if $|x - x_0| < \delta$. Now let $0 < |\Delta x| < \delta$. By Theorem 2.6 there exists some ξ strictly between x_0 and $x_0 + \Delta x$ such that

$$f(x_0 + \Delta x) = f(x_0) + f'(x_0)\Delta x + \tfrac{1}{2}f''(\xi)(\Delta x)^2 . \tag{2.13}$$

Since $f'(x_0) = 0$ and $f''(\xi) > 0$ the result follows by inspection. Note that if we assume $f'(x_0) = 0$ and $f''(x_0) < 0$, then f has a local maximum at x_0. We will state and prove the theorem for n variables later. □

Differentiation is a handy device for checking many proposed inequalities. One plan is as follows. Suppose the proposed inequality is of the general form

$$g(x) < h(x) \qquad (x > x_0), \qquad (2.14)$$

where $g(x_0) = h(x_0)$ and the functions $g(x)$ and $h(x)$ have known derivatives. Defining $f(x) = h(x) - g(x)$, we have $f(x_0) = 0$. If we can further show that $f'(x) > 0$ for $x > x_0$, then (2.14) is established.

Example 2.10. We can prove that for $x > 0$ we have

$$x^r \le rx + (1 - r) \qquad (0 < r < 1). \qquad (2.15)$$

Defining $f(x) = (1 - r) + rx - x^r$, we find $f(1) = 0$ and

$$f'(x) = r - rx^{r-1} = r\left(1 - \frac{1}{x^{1-r}}\right).$$

For $x > 1$ we have $f'(x) > 0$; for $0 < x < 1$ we have $f'(x) < 0$. Hence (2.15) holds, with equality if and only if $x = 1$. Similarly, for $x > 0$ we have

$$x^r \ge rx + (1 - r) \qquad (r < 0 \text{ or } r > 1). \qquad (2.16)$$

Beckenbach and Bellman [7] call these inequalities "remarkable" as they can be used to derive the AM–GM, Hölder, and Minkowski inequalities of Chap. 3.   □

Example 2.11. For $0 < x < \pi/2$, relation (2.10) yields

$$\frac{d}{dx}\left(\frac{\sin x}{x}\right) = \cos x\left(\frac{x - \tan x}{x^2}\right) < 0$$

so $\sin x/x$ is strictly decreasing. Because $\sin x/x \to 2/\pi$ as $x \to \pi/2$, we conclude that

$$\sin x > 2x/\pi \qquad (0 < x < \pi/2). \qquad (2.17)$$

This is *Jordan's inequality* (Fig. 2.3). The role of concavity suggested here will be exploited further in Sect. 3.9.   □

Inequalities are often obtained by solving constrained optimization problems via the Lagrange multiplier technique. The main idea is as follows. To prove an inequality of the form

$$f(x, y) \le g(x, y), \qquad (2.18)$$

we can try to maximize the function $f = f(x, y)$ subject to the condition $g(x, y) = k$ where k is a constant. If for any permissible k the constrained maximum value of $f(x, y)$ is f_{max} and if $f_{max} \le k$, then (2.18) is proved.

Alternatively, we could try to minimize the right member $g(x, y)$ subject to the condition $f(x, y) = c$ with c a constant. If for any permissible c the constrained minimum value of $g(x, y)$ is g_{min} and if $g_{min} \ge c$, then (2.18) is likewise established. Let us carry out this procedure to prove a standard inequality obtained (by a different

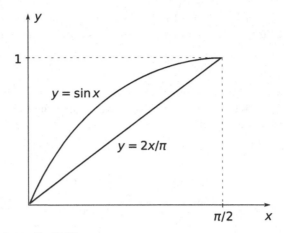

Fig. 2.3 Jordan's inequality (2.17)

approach) in the next chapter. The desired result is a special case of *Young's in-equality*. It states that if p and q are positive numbers for which $p^{-1} + q^{-1} = 1$, then the inequality

$$xy \leq \frac{x^p}{p} + \frac{y^q}{q} \tag{2.19}$$

holds for all nonnegative numbers x and y.

We will minimize the right member

$$g(x, y) = \frac{x^p}{p} + \frac{y^q}{q} \tag{2.20}$$

subject to the constraint that the left member

$$f(x, y) = xy = c, \text{ a constant.} \tag{2.21}$$

Because (2.19) holds trivially when either x or y is zero, we can assume $c \neq 0$. Carrying out the usual Lagrange multiplier technique, we form the Lagrangian function with a multiplier $-\lambda$ (negative sign arbitrary but taken for convenience) as

$$F(x, y) = \frac{x^p}{p} + \frac{y^q}{q} - \lambda xy$$

and differentiate this function with respect to x and y, respectively, to obtain the equations

$$x^{p-1} - \lambda y = 0, \tag{2.22}$$

$$y^{q-1} - \lambda x = 0. \tag{2.23}$$

We then solve the system consisting of (2.22)–(2.23) and the constraint (2.21). The solution is straightforward; we find a constrained stationary point for $g(x, y)$ at

$$(x_s, y_s) = (c^{1/p}, c^{1/q}) .$$

Furthermore, $g(x_s, y_s) = c$. Because $g(x, y)$ in (2.20) is not bounded above when subject to a constraint of the form (2.21), it is clear that (x_s, y_s) is actually the constrained minimum of $g(x, y)$ corresponding to a given value of c.

Finally, any point (x, y) with $x > 0$ and $y > 0$ has a hyperbola of the form $xy = c$ passing through it. The relation

$$g(x, y) = \frac{x^p}{p} + \frac{y^q}{q} \geq g_{min} = c = xy$$

gives Young's inequality (2.19).

2.5 Problems

2.1. The following exercises involve monotonicity.

(a)   Show that if $n \in \mathbb{N}$ then

$$\ln(n + 1) > \frac{1}{n} \sum_{k=1}^{n} \ln k .$$

(b)   Show that if ϕ, ψ, and f are increasing functions with $\phi(x) \leq f(x) \leq \psi(x)$, then

$$\phi(\phi(x)) \leq f(f(x)) \leq \psi(\psi(x)) .$$

(c)   [56, 65] Show that if f is increasing on $[a, b]$, then

$$\frac{1}{x - a} \int_a^x f(u) \, du \leq \frac{1}{b - a} \int_a^b f(u) \, du \leq \frac{1}{b - x} \int_x^b f(u) \, du \qquad (2.24)$$

for any $x \in (a, b)$.

2.2. The following exercises involve the definition of integration. Recall from calculus that f is integrable on $[a, b]$ and

$$\int_a^b f(x) \, dx = I$$

means that given $\varepsilon > 0$ there exists some $\delta > 0$ such that if

$$a = x_0 < x_1 < \cdots < x_n = b$$

and if $x_i - x_{i-1} < \delta$ for $i = 1, \ldots, n$ and if $\xi_i \in [x_{i-1}, x_i]$ for $i = 1, \ldots, n$, then

$$\left| \sum_{i=1}^{n} f(\xi_i)(x_i - x_{i-1}) - I \right| < \delta .$$

Note as a special case that if f is integrable on $[a, b]$, then

$$\lim_{n\to\infty} \sum_{i=1}^{n} f(a + i\Delta x)\Delta x = \int_a^b f(x)\, dx \text{ where } \Delta x = (b-a)/n .$$

(a)   Show that if f is integrable on $[a, b]$, then f is bounded on $[a, b]$.

(b)   Show that if f is integrable on $[a, b]$, then F given by

$$F(x) = \int_a^x f(t)\, dt$$

is continuous on $[a, b]$.

(c)   Define $f(x) = x^{-1/2}$ if $0 < x \le 1$ and $f(0) = 0$. Does $\int_0^1 f(x)\, dx$ exist?

2.3. The exercises below also involve integration.

(a)   Put simple lower and upper bounds on the family of integrals

$$I(\alpha, \beta) = \int_0^1 \frac{dx}{(x^\beta + 1)^\alpha} \qquad (\alpha, \beta \ge 0) .$$

(b)   Show that

$$\int_0^{\pi/2} \ln(1/\sin t)\, dt < \infty .$$

(c)   A function f is *of exponential order* on $[0, \infty)$ if there exist positive numbers b and C such that $|f(t)| \le Ce^{bt}$ for $t \ge 0$. Show that the Laplace transform of f given by

$$F(s) = \int_0^\infty f(t)e^{-st}\, dt$$

exists if f is of exponential order.

(d)   Verify that

$$\int_0^{\pi/2} (\sin x)^{2n+1}\, dx \le \int_0^{\pi/2} (\sin x)^{2n}\, dx \le \int_0^{\pi/2} (\sin x)^{2n-1}\, dx$$

and establish *Wallis's product*

$$\frac{\pi}{2} = \frac{2}{1}\cdot\frac{2}{3}\cdot\frac{4}{3}\cdot\frac{4}{5}\cdot\frac{6}{5}\cdot\frac{6}{7}\cdots\cdots\frac{2m}{2m-1}\cdot\frac{2m}{2m+1}\cdots\cdots .$$

(e)   Show that

$$\lim_{T\to\infty} \int_0^T \frac{\sin x}{x}\, dx$$

exists and is between 1 and 3.

(f)   Prove that if g is continuous on $[a, b]$ with $g(x) \ge 0$ and $\int_a^b g(x)\, dx = 0$, then $g(x) \equiv 0$ on $[a, b]$.

(g)   Let $p \in \mathbb{R}$, $p > 0$. Use the fact that $\ln x = \int_1^x dt/t$ and the squeeze principle to show that

$$\lim_{x\to\infty} (\ln x)/(x^p) = 0 .$$

2.4. Let f and g be functions integrable on (a, b), with $0 \le g(t) \le 1$ and f decreasing on (a, b). Prove *Steffensen's inequality* [7, 56]

$$\int_{b-\lambda}^{b} f(t)\,dt \le \int_{a}^{b} f(t)g(t)\,dt \le \int_{a}^{a+\lambda} f(t)\,dt \quad \text{where } \lambda = \int_{a}^{b} g(t)\,dt \ .$$

2.5. Prove the following statements. Parts (a) and (b) are challenging; according to Hobson [37], they were first given by Bonnet circa 1850.

(a)   Let f be a monotonic decreasing, nonnegative function on $[a, b]$, and let g be integrable on $[a, b]$. Then for some ξ with $a \le \xi \le b$,

$$\int_{a}^{b} f(x)g(x)\,dx = f(a) \int_{a}^{\xi} g(x)\,dx \ .$$

(b)   Let f be a monotonic increasing, nonnegative function on $[a, b]$, and let g be integrable on $[a, b]$. Then for some η with $a \le \eta \le b$,

$$\int_{a}^{b} f(x)g(x)\,dx = f(b) \int_{\eta}^{b} g(x)\,dx \ .$$

(c)   Let f be bounded and monotonic on $[a, b]$, and let g be integrable on $[a, b]$. Then for some ξ with $a \le \xi \le b$,

$$\int_{a}^{b} f(x)g(x)\,dx = f(a) \int_{a}^{\xi} g(x)\,dx + f(b) \int_{\xi}^{b} g(x)\,dx \ .$$

This is also called the second mean value theorem for integrals, particularly in older books.

(d)   Let f be a monotonic function integrable on $[a, b]$, and suppose that $f(a)f(b) \ge 0$ and $|f(a)| \ge |f(b)|$. Then, if g is a real function integrable on $[a, b]$,

$$\left| \int_{a}^{b} f(x)g(x)\,dx \right| \le |f(a)| \cdot \max_{a \le \xi \le b} \left| \int_{a}^{\xi} g(x)\,dx \right| \ .$$

This is *Ostrowski's inequality for integrals*.

2.6. Use graphical approaches to complete the following.

(a)   Show that if f is increasing on $[0, \infty)$, then

$$\int_{0}^{n} f(x)\,dx \le \sum_{k=1}^{n} f(k) \le \int_{1}^{n+1} f(x)\,dx \ .$$

Use this to find upper and lower bounds on $\sum_{k=1}^{n} k^2$.

(b)   Show that

$$\int_{1}^{n} \ln x\,dx < \ln(n!) < \int_{1}^{n+1} \ln x\,dx \qquad (n \in \mathbb{N}, n > 1) \ .$$

(c)   Sketch the curve $y = 1/x$ for $x > 0$, and consider the area bounded by this curve, the x-axis, and the lines $x = a$ and $x = b$ ($b > a$). Compare this with the areas of two trapezoids and obtain

$$\frac{2(b-a)}{b+a} < \ln \frac{b}{a} < \frac{b^2 - a^2}{2ab} \ .$$

(d)   (*Integral test inequality* [13].) Show that if f is decreasing on $[1, \infty)$ then

$$\sum_{k=2}^{n} f(k) \le \int_1^n f(x)\,dx \le \sum_{k=1}^{n-1} f(k).$$

(e)   Show that if f is increasing on $[1, \infty)$, then

$$\sum_{k=1}^{n-1} f(k) \le \int_1^n f(x)\,dx \le \sum_{k=2}^{n} f(k).$$

(f)   Show that

$$1 + \frac{1}{2^3} + \frac{1}{3^3} + \cdots + \frac{1}{n^3} < \frac{5}{4}.$$

(g)   *Euler's constant* C is defined by

$$C = \lim_{n \to \infty} C_n = \lim_{n \to \infty} \left(\sum_{j=1}^{n} \frac{1}{j} - \ln n \right).$$

Verify that C exists and is positive by showing that C_n is strictly decreasing with lower bound $1/2$.

2.7. Use differentiation to prove the results below [1, 57, 69]. Assume $n \in \mathbb{N}$.

(a) $\ln x \le x - 1$ for $x > 0$, with equality if and only if $x = 1$,

(b) $\ln x \le n(x^{1/n} - 1)$ for $x > 0$, with equality if and only if $x = 1$,

(c) $x^n + (n - 1) \ge x$ for $x \ge 0$,

(d) $2\ln(\sec x) < \sin x \tan x$ for $0 < x < \pi/2$,

(e) $\sinh x \ge x$ for $x \ge 0$,

(f) $|x \ln x| \le e^{-1}$ for $0 \le x \le 1$,

(g) $e^x < (1 - x)^{-1}$ for $x < 1$ and $x \ne 0$,

(h) $\pi^e < e^\pi$ (more generally [13], $e^x > x^e$ for any $x \ne e$),

(i) $(s + t)^a \le s^a + t^a \le 2^{1-a}(s + t)^a$ for $s, t > 0$ and $0 < a \le 1$,

(j) $2^{1-b}(s + t)^b \le s^b + t^b \le (s + t)^b$ for $s, t > 0$ and $b \ge 1$,

(k) $e^x \ge (ex/a)^a$ for $x > a$ and $a > 0$,

(l) $x^x \ge e^{x-1}$ for $x > 0$,

(m) $\cos x \ge 1 - x^2/2$ for $x \ge 0$,

(n) $\sin x \ge x - x^3/3!$ for $x \ge 0$.

2.8. Use Corollary 2.4 to derive the following inequalities [57, 69]:

(a) $\sin x < x$ for $x > 0$,

(b) $x/(1 + x^2) < \tan^{-1} x < x$ for $x > 0$,

(c) $1 + (x/2\sqrt{1 + x}) < \sqrt{1 + x} < 1 + x/2$ for $x > 0$,

(d) $e^x(y - x) < e^y - e^x < e^y(y - x)$ for $y > x$,

(e) $(1 + x)^a \le 1 + ax(1 + x)^{a-1}$ for $a > 1$ and $x > -1$, with equality if and only if $x = 0$,

(f) $1 + x > e^{x/(1+x)}$ for $x > -1$ and $x \ne 0$,

(g) $(y - x)/\cos^2 x < \tan y - \tan x < (y - x)/\cos^2 y$ for $0 \le x < y < \pi/2$,

(h) $ex < (y^y/x^x)^{1/(y-x)} < ey$ for $0 < x < y$.

2.9. Derive a nonstrict version of (2.10) by integration.

2.10. The following are applications of l'Hôpital's monotone rule.

(a)   For $a > 1$ and $x > -1$, $x \neq 0$, define

$$h(x) = \frac{(1 + x)^a - 1}{x} .$$

Use l'Hôpital's rule to define $h(0) = a$. Use Theorem 2.9 to show that $h(x)$ is increasing on $[-1, \infty)$ and hence that

$$(1 + x)^a \geq 1 + ax$$

with equality if and only if $x = 0$ (cf., Example 2.10).

(b)   Show that

$$h(x) = \frac{\ln \cosh x}{\ln((\sinh x)/x)}$$

is decreasing on $(0, \infty)$.

(c)   Prove that for $x \in (0, 1)$,

$$\pi < \frac{\sin \pi x}{x(1 - x)} \leq 4 .$$

(d)   Prove that $1 > \sin x/x > 2/\pi$ on $(0, \pi/2)$ (cf., Example 2.11.)

2.11. Use series expansions to establish the following inequalities:

(a) $|\cos z| \leq \cosh |z|$ for $z \in \mathbb{C}$,

(b) $|\ln(1 + x)| \leq -\ln(1 - |x|)$ if $|x| < 1$,

(c) $\prod_{n=1}^{\infty}(1 + a_n) \leq \exp\left(\sum_{n=1}^{\infty} a_n\right)$ if $0 \leq a_n < 1$ for all n,

(d) $e^x > 1 + x^n/n!$ for $n \in \mathbb{N}$ and $x > 0$,

(e) $x < e^x - 1 < x/(1 - x)$ for $x < 1$ and $x \neq 0$.

2.12. Show that if n is an integer greater than 1 and a, b are positive with $a > b$, then

$$b^{n-1} < \frac{a^n - b^n}{n(a - b)} < a^{n-1} .$$

Use this to prove that no positive real number can have more than one positive nth root.

2.13. Prove the following generalized version of Rolle's theorem. Let g be n times continuously differentiable on $[a, b]$, and let $x_0 < x_1 < \cdots < x_n$ be $n+1$ points in $[a, b]$. Suppose $g(x_0) = g(x_1) = \cdots = g(x_n) = 0$. Then there exists $\xi \in [a, b]$ such that $g^{(n)}(\xi) = 0$.

2.14. A set A is said to be *dense* in a set B if every element of B is the limit of a sequence of elements belonging to A. Show that if $f(x)$ and $g(x)$ are continuous on B with $f(x) \leq g(x)$ for every x in some dense subset of B, then $f(x) \leq g(x)$ for all $x \in B$. Explain how this idea could be used to extend to real arguments an inequality proved for rational arguments.

2.15. (A simple caution.) Given a valid inequality between two functions, is it generally possible to obtain another valid inequality by direct differentiation? Is it true, for instance, that $f'(x) > g'(x)$ whenever $f(x) > g(x)$? Note, however, that if $f'(x) > g'(x)$ on $[a, b]$, then we do have $f(b) - f(a) > g(b) - g(a)$.

2.16. Use Lagrange multipliers to show that

$$\frac{x^n + y^n}{2} \geq \left(\frac{x + y}{2}\right)^n$$

for $n \geq 1$ and $x, y \geq 0$.

Chapter 3
Some Standard Inequalities

3.1 Introduction

Here we examine certain famous inequalities that have left bold imprints on both pure and applied mathematics. These results, some of which are very old, pertain to functions, sequences, and integrals. We recall that integral inequalities are frequently deduced by establishing the corresponding result for series, writing it out for Riemann sums, and then implementing a limit passage. However, this is not the only method by which integral inequalities can be obtained.

Classic reference books for the material of this chapter include [6, 34, 56].

3.2 Bernoulli's Inequality

Theorem 3.1 (Bernoulli's Inequality). *If $n \in \mathbb{N}$ and $x \geq -1$, then*

$$(1 + x)^n \geq 1 + nx . \tag{3.1}$$

Equality holds if and only if $n = 1$ or $x = 0$.

Proof. We can give a simple proof by induction. Let $\mathcal{P}(n)$ be the proposition

$$x \geq -1 \implies (1 + x)^n \geq 1 + nx \text{ with equality if and only if } n = 1 \text{ or } x = 0 .$$

The case $\mathcal{P}(1)$ holds trivially. Now let $n \in \mathbb{N}$ and assume $\mathcal{P}(n)$ is true. Note that since $n + 1 \neq 1$, conditions for equality in $\mathcal{P}(n + 1)$ are simply $x = 0$. Multiplying by the nonnegative number $1 + x$, we have

$$(1 + x)^{n+1} \geq (1 + x)(1 + nx) = 1 + (n + 1)x + nx^2 \geq 1 + (n + 1)x . \tag{3.2}$$

Equality holds in (3.2) if and only if $nx^2 = 0$, which holds if and only if $x = 0$. $\quad\square$

See Problem 2.10 for a generalization.

M.J. Cloud et al., *Inequalities: With Applications to Engineering*,
DOI 10.1007/978-3-319-05311-0_3, © Springer International Publishing AG 2014

3.3 Young's Inequality

Consider two continuous functions f and g, both strictly increasing and inverses of each other. Suppose the functions vanish at the origin as in Fig. 3.1. Area $A + B$ clearly exceeds the area of a rectangle of width w and height h (for any choice of positive numbers w, h), and we are led immediately to the following theorem.

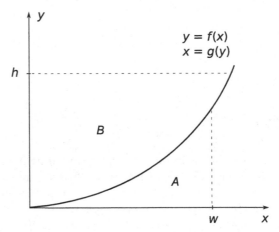

Fig. 3.1 Young's inequality (3.3)

Theorem 3.2 (Young's Inequality). *Let f, g be continuous, strictly increasing, and mutually inverse for nonnegative argument, with $f(0) = g(0) = 0$. Then*

$$wh \leq \int_0^w f(x)\,dx + \int_0^h g(x)\,dx \,. \tag{3.3}$$

Equality holds if and only if $h = f(w)$.

Analytical proofs can be found in [24, 56].

3.4 Inequality of the Means

We now present the celebrated arithmetic mean–geometric mean, or AM–GM, inequality.

Theorem 3.3 (Weighted AM–GM Inequality). *Let $a_1, \ldots, a_n$ be positive numbers and let $\delta_1, \ldots, \delta_n$ be positive numbers (weights) such that $\delta_1 + \cdots + \delta_n = 1$. Then*

$$\delta_1 a_1 + \cdots + \delta_n a_n \geq a_1^{\delta_1} \cdots a_n^{\delta_n} \tag{3.4}$$

and equality holds if and only if the a_i are all equal.

Proof. (See [22]). We begin with the fact that $x - 1 - \ln x \geq 0$ whenever $x > 0$, with equality if and only if $x = 1$ (Problem 2.7). Call

$$A = \sum_{k=1}^{n} \delta_k a_k \ .$$

For each i, we have $a_i/A - 1 - \ln(a_i/A) \geq 0$. Multiplying each such term by δ_i and summing over i, we get

$$\sum_{i=1}^{n} (\delta_i a_i/A - \delta_i) - \sum_{i=1}^{n} \delta_i \ln(a_i/A) \geq 0 \ . \tag{3.5}$$

Since the first summation vanishes, we have

$$\sum_{i=1}^{n} \delta_i \ln(a_i/A) \leq 0 \ .$$

Because the exponential function is increasing,

$$\exp\left[\sum_{i=1}^{n} \delta_i \ln(a_i/A) \right] \leq \exp(0) = 1 \ .$$

Hence $(a_1^{\delta_1} \cdots a_n^{\delta_n})/A \leq 1$, and

$$a_1^{\delta_1} \cdots a_n^{\delta_n} \leq \delta_1 a_1 + \cdots + \delta_n a_n \ . \tag{3.6}$$

Equality holds in (3.6) if and only if it holds in (3.5). Because each summand is nonnegative, equality holds in (3.5) if and only if each summand is zero which is equivalent to each $a_i/A = 1$. In other words, equality holds in (3.6) if and only if $a_1 = \cdots = a_n$.

For other proofs, see Problems 3.6 and 3.7. $\qquad\qquad\qquad\qquad\qquad\square$

The choice of weights $\delta_i = 1/n$ for all i leads to the next result.

Corollary 3.1 (AM–GM Inequality). *If $a_1, \ldots, a_n$ are positive numbers, then*

$$\frac{a_1 + \cdots + a_n}{n} \geq (a_1 \cdots a_n)^{1/n} \ . \tag{3.7}$$

Equality holds if and only if the a_i are all equal.

The left member is the ordinary arithmetic mean of the n numbers, while the right member is by definition the ordinary geometric mean. Note that for given positive numbers a_i, the AM–GM inequality (3.7) provides a lower bound for a sum as

$$a_1 + \cdots + a_n \geq n(a_1 \cdots a_n)^{1/n} \ ,$$

or an upper bound for a product as

$$a_1 \cdots a_n \le \left(\frac{a_1 + \cdots + a_n}{n}\right)^n .$$

Example 3.1. For any real x and y, the numbers e^x and e^y are positive. Therefore

$$\frac{e^x + e^y}{2} \ge e^{(x+y)/2}$$

with equality if and only if $x = y$.   □

Example 3.2. Application of (3.7) to the reciprocals $1/a_i$ gives

$$\frac{n}{a_1^{-1} + \cdots + a_n^{-1}} \le (a_1 \cdots a_n)^{1/n} .$$

The left member is the *harmonic mean* of the a_i. Thus the harmonic mean of positive numbers never exceeds the geometric mean. This result can be extended to the inequality

$$\min_{1 \le i \le n} a_i \le \frac{n}{\sum_{i=1}^{n} \frac{1}{a_i}} \le \left(\prod_{i=1}^{n} a_i\right)^{1/n} \le \frac{\sum_{i=1}^{n} a_i}{n} \le \left(\frac{\sum_{i=1}^{n} a_i^2}{n}\right)^{1/2} \le \max_{1 \le i \le n} a_i \qquad (3.8)$$

for positive numbers $a_1, \ldots, a_n$. Equality holds if and only if $a_1 = a_2 = \cdots = a_n$. The second term from the right in (3.8) is the *quadratic mean* (or root-mean-square value) of the numbers a_i.   □

Example 3.3. A simple technique is to multiply by unity and then apply (3.7). Consider, for instance, the sequence $\{a_n\}$ with $a_n = (1 + 1/n)^n$. We have

$$a_n = \left(1 + \frac{1}{n}\right)^n \cdot 1 < \left[\frac{n\left(1 + \frac{1}{n}\right) + 1}{n + 1}\right]^{n+1} = \left(\frac{n + 2}{n + 1}\right)^{n+1} = \left(1 + \frac{1}{n + 1}\right)^{n+1} = a_{n+1} .$$

Hence $\{a_n\}$ is strictly increasing.   □

An integral form of the AM–GM inequality is introduced in Problem 3.8.

3.5 Hölder's Inequality

This result can be obtained in one step from the weighted AM–GM inequality.

Theorem 3.4 (Hölder's Inequality). *Suppose for each j, $1 \le j \le n$, that $a_{j1}, \ldots, a_{jm}$ are nonzero numbers. Suppose $\delta_1, \ldots, \delta_n$ are positive numbers such that $\delta_1 + \cdots + \delta_n = 1$. For each j denote*

$$S_j = \sum_{i=1}^{m} |a_{ji}| \; .$$

Then

$$\sum_{i=1}^{m} |a_{1i}|^{\delta_1} \cdots |a_{ni}|^{\delta_n} \le S_1^{\delta_1} \cdots S_n^{\delta_n} \; . \tag{3.9}$$

Proof.

$$\frac{\sum_{i=1}^{m} |a_{1i}|^{\delta_1} \cdots |a_{ni}|^{\delta_n}}{S_1^{\delta_1} \cdots S_n^{\delta_n}} = \sum_{i=1}^{m} \left(\frac{|a_{1i}|}{S_1}\right)^{\delta_1} \cdots \left(\frac{|a_{ni}|}{S_n}\right)^{\delta_n} \le \sum_{i=1}^{m} \left(\delta_1 \frac{|a_{1i}|}{S_1} + \cdots + \delta_n \frac{|a_{ni}|}{S_n}\right)$$

$$= \delta_1 + \cdots + \delta_n = 1 \tag{3.10}$$

by the application of (3.4) to each summand. □

With $n = 2$ write $\delta_1 = 1/p$, $\delta_2 = 1/q$, and let $a_{1i} = |a_i|^p$ and $a_{2i} = |b_i|^q$ for $i = 1, \ldots, m$. Then (3.9) becomes

$$\sum_{i=1}^{m} |a_i b_i| \le \left(\sum_{i=1}^{m} |a_i|^p\right)^{1/p} \left(\sum_{i=1}^{m} |b_i|^q\right)^{1/q} \; . \tag{3.11}$$

This special case is also commonly referred to as Hölder's inequality, and we can give another proof based on Young's inequality. Putting $f(x) = x^{p-1}$ and $g(x) = x^{q-1}$ with

$$\frac{1}{p} + \frac{1}{q} = 1 \qquad (1 < p < \infty) \tag{3.12}$$

we obtain from (3.3)

$$wh \le \frac{w^p}{p} + \frac{h^q}{q} \; . \tag{3.13}$$

With two sets of m numbers $a_1, \ldots, a_m$ and $b_1, \ldots, b_m$, we form the quantities

$$\alpha = \left(\sum_{j=1}^{m} |a_j|^p\right)^{1/p}, \qquad \beta = \left(\sum_{j=1}^{m} |b_j|^q\right)^{1/q} \; .$$

Assuming that α, β are both nonzero, from (3.13) we have

$$\frac{|a_i|}{\alpha} \frac{|b_i|}{\beta} \le \frac{1}{p} \frac{|a_i|^p}{\alpha^p} + \frac{1}{q} \frac{|b_i|^q}{\beta^q}$$

for any positive integer i. Summation over i produces

$$\frac{1}{\alpha\beta} \sum_{i=1}^{m} |a_i||b_i| \le \frac{1}{p\alpha^p} \sum_{i=1}^{m} |a_i|^p + \frac{1}{q\beta^q} \sum_{i=1}^{m} |b_i|^q = \frac{1}{p} + \frac{1}{q} = 1 \; ,$$

as required.

Taking $m \to \infty$ we have, by Lemma 2.1,

$$\sum_{i=1}^{\infty} |a_i b_i| \le \left(\sum_{i=1}^{\infty} |a_i|^p \right)^{1/p} \left(\sum_{i=1}^{\infty} |b_i|^q \right)^{1/q}$$

provided the series on the right both converge. The corresponding result for integrals, if the integrals on the right side exist, implies that the integral on the left side also exists and

$$\int_a^b |f(x) g(x)| \, dx \le \left(\int_a^b |f(x)|^p \, dx \right)^{1/p} \left(\int_a^b |g(x)|^q \, dx \right)^{1/q}. \qquad (3.14)$$

See Problem 3.14 for a derivation.

In order to discuss when equality holds in Hölder's inequality, we note that if $\alpha_i \ge 0$ for all i, then

$$\sum_{i=1}^{m} \alpha_i = 0$$

if and only if each $\alpha_i = 0$. If $\alpha_i \ge \beta_i$ for all i, then

$$\sum_{i=1}^{m} \alpha_i = \sum_{i=1}^{m} \beta_i$$

if and only if $\alpha_i = \beta_i$ for all i. Thus equality holds in (3.10) if and only if for each i

$$\left(\frac{|a_{1i}|}{S_1} \right)^{\delta_1} \cdots \left(\frac{|a_{ni}|}{S_n} \right)^{\delta_n} = \delta_1 \frac{|a_{1i}|}{S_1} + \cdots + \delta_n \frac{|a_{ni}|}{S_n}. \qquad (3.15)$$

From the weighted AM–GM inequality, (3.15) holds for each i if and only if

$$\frac{|a_{1i}|}{S_1} = \cdots = \frac{|a_{ni}|}{S_n}. \qquad (3.16)$$

Hence equality holds in Hölder's inequality (3.9) if and only if (3.16) holds for all i. In the case $n = 2$, equality holds in (3.11) if and only if

$$\frac{|a_i|^p}{\sum_{i=1}^{m} |a_i|^p} = \frac{|b_i|^q}{\sum_{i=1}^{m} |b_i|^q} \qquad (3.17)$$

for all i.

It is convenient to remove the condition that each a_{ji} be nonzero. If $a_{j1} = \cdots = a_{jm} = 0$, then (3.9) holds by inspection. Now suppose each set $\{a_{j1}, \ldots, a_{jm}\}$ contains at least one nonzero term. For each index i in (3.10) it is still true that

$$\left(\frac{|a_{1i}|}{S_1} \right)^{\delta_1} \cdots \left(\frac{|a_{ni}|}{S_n} \right)^{\delta_n} \le \delta_1 \frac{|a_{1i}|}{S_1} + \cdots + \delta_n \frac{|a_{ni}|}{S_n}$$

(by (3.4) if each $a_{ji} \ne 0$, by inspection otherwise) so (3.10) and (3.11) are still valid. We summarize this discussion applied to (3.11) as follows:

Theorem 3.5 (Hölder's Inequality). *Let $p > 1$ and $q > 1$ and $p^{-1} + q^{-1} = 1$. Let $a_1, \ldots, a_m$ and $b_1, \ldots, b_m$ be two sequences of real numbers. Then*

$$\sum_{i=1}^{m} |a_i b_i| \leq \left(\sum_{i=1}^{m} |a_i|^p \right)^{1/p} \left(\sum_{i=1}^{m} |b_i|^q \right)^{1/q}. \tag{3.18}$$

Equality holds if and only if one of the sequences a_i or b_i consists entirely of zeros or else

$$\frac{|a_i|^p}{\sum_{i=1}^{m} |a_i|^p} = \frac{|b_i|^q}{\sum_{i=1}^{m} |b_i|^q} \quad \text{for all } i. \tag{3.19}$$

Problem 3.12 gives an equivalent way to state the condition for equality.

Remark 3.1. Exponents satisfying (3.12) are known as *complementary* or *conjugate exponents*. Other ways to state (3.12) include

$$(p - 1)(q - 1) = 1, \qquad q = \frac{p}{p - 1}, \qquad p = \frac{q}{q - 1}, \qquad p + q = pq.$$

Also notice that if $p = 2$ then $q = 2$, and if $p > 2$ then $q < 2$. $\qquad\square$

3.6 Minkowski's Inequality

Theorem 3.6 (Minkowski's Inequality). *Assume that $a_1, \ldots, a_m$ and $b_1, \ldots, b_m$ are real numbers, and let $p \geq 1$. Then*

$$\left(\sum_{i=1}^{m} |a_i + b_i|^p \right)^{1/p} \leq \left(\sum_{i=1}^{m} |a_i|^p \right)^{1/p} + \left(\sum_{i=1}^{m} |b_i|^p \right)^{1/p}. \tag{3.20}$$

Proof. If $p - 1$ this follows from the triangle inequality. Now suppose $p > 1$, and choose $q > 1$ so that $p^{-1} + q^{-1} = 1$. Write Hölder's inequality as

$$\sum_{i=1}^{m} |\alpha_i \beta_i| \leq \left(\sum_{i=1}^{m} |\alpha_i|^p \right)^{1/p} \left(\sum_{i=1}^{m} |\beta_i|^q \right)^{1/q}.$$

Let $\alpha_i = |a_i|$ and $\beta_i = |a_i + b_i|^{p/q}$, and then let $\alpha_i = |b_i|$ and $\beta_i = |a_i + b_i|^{p/q}$ to get

$$\sum_{i=1}^{m} |a_i| \, |a_i + b_i|^{p/q} \leq \left(\sum_{i=1}^{m} |a_i|^p \right)^{1/p} \left(\sum_{i=1}^{m} |a_i + b_i|^p \right)^{1/q} \tag{3.21}$$

and

$$\sum_{i=1}^{m} |b_i| \, |a_i + b_i|^{p/q} \leq \left(\sum_{i=1}^{m} |b_i|^p \right)^{1/p} \left(\sum_{i=1}^{m} |a_i + b_i|^p \right)^{1/q}, \tag{3.22}$$

respectively. Since $p = 1 + (p/q)$, for each i,

$$|a_i + b_i|^p = |a_i + b_i| \, |a_i + b_i|^{p/q} \leq |a_i| \, |a_i + b_i|^{p/q} + |b_i| \, |a_i + b_i|^{p/q} . \qquad (3.23)$$

Summing over the terms in (3.23), and using (3.22) and (3.21), we have

$$\sum_{i=1}^{m} |a_i + b_i|^p \leq \left(\sum_{i=1}^{m} |a_i|^p \right)^{1/p} \left(\sum_{i=1}^{m} |a_i + b_i|^p \right)^{1/q} + \left(\sum_{i=1}^{m} |b_i|^p \right)^{1/p} \left(\sum_{i=1}^{m} |a_i + b_i|^p \right)^{1/q} .$$

We may assume that $\sum_{i=1}^{m} |a_i + b_i|^p \neq 0$ (because (3.20) obviously holds otherwise). Hence we complete the proof by dividing through by $(\sum_{i=1}^{m} |a_i + b_i|^p)^{1/q}$ and using the fact that $1 - 1/q = 1/p$.   □

For conditions when equality holds, see Problem 4.8.

Minkowski's inequality can be extended to infinite series, provided the series converge, as

$$\left(\sum_{i=1}^{\infty} |a_i + b_i|^p \right)^{1/p} \leq \left(\sum_{i=1}^{\infty} |a_i|^p \right)^{1/p} + \left(\sum_{i=1}^{\infty} |b_i|^p \right)^{1/p}$$

and to integrals, provided the integrals exist, as

$$\left(\int_a^b |f(x) + g(x)|^p \, dx \right)^{1/p} \leq \left(\int_a^b |f(x)|^p \, dx \right)^{1/p} + \left(\int_a^b |g(x)|^p \, dx \right)^{1/p} .$$

Minkowski's inequality is a statement of the fact that for a vector $\mathbf{a} = (a_1, \ldots, a_n)$, the expression

$$\|\mathbf{a}\| = \left(\sum_{i=1}^{\infty} |b_i|^p \right)^{1/p}$$

possesses norm property N_3 on p. 21. Clearly properties N_1 and N_2 hold as well, and so for $p \geq 1$ the expression $\|\cdot\|$ is a norm on $\mathbb{R}^n$. As all norms are equivalent on $\mathbb{R}^n$, we may establish a variety of inequalities between various norms on $\mathbb{R}^n$, including Minkowski's inequality.

3.7 Cauchy–Schwarz Inequality

Theorem 3.7 (Cauchy–Schwarz Inequality). *Suppose $a_1, \ldots, a_m$ and $b_1, \ldots, b_m$ are real numbers. Then*

$$\left| \sum_{i=1}^{m} a_i b_i \right| \leq \left(\sum_{i=1}^{m} a_i^2 \right)^{1/2} \left(\sum_{i=1}^{m} b_i^2 \right)^{1/2} \qquad (3.24)$$

or equivalently

$$\left(\sum_{i=1}^{m} a_i b_i \right)^2 \le \sum_{i=1}^{m} a_i^2 \sum_{i=1}^{m} b_i^2 . \tag{3.25}$$

Equality holds if and only if all a_i are zero, or all b_i are zero, or there exists $\lambda \in \mathbb{R}$ such that $a_i = \lambda b_i$ for all i.

Proof. Take $\mathbf{x} = \mathbf{a} = (a_1, \ldots, a_n)$, $\mathbf{y} = \mathbf{b} = (b_1, \ldots, b_n)$, and $\langle \mathbf{a}, \mathbf{b} \rangle = a_1 b_1 + \cdots + a_n b_n$ in the general version of the Cauchy–Schwarz inequality (1.24). $\qquad\square$

The Cauchy–Schwarz inequality can also be regarded as a special case of Hölder's inequality for $p = q = 2$. The form (3.25) is so important that a good portion of the book [6] is devoted to many versions of its proof.

Provided the series on the right converge, (3.25) yields

$$\left(\sum_{i=1}^{\infty} a_i b_i \right)^2 \le \sum_{i=1}^{\infty} a_i^2 \sum_{i=1}^{\infty} b_i^2 .$$

We can also write (3.25) for Riemann sums and apply Lemma 2.1 to obtain

$$\left(\int_a^b f(x) g(x)\, dx \right)^2 \le \int_a^b f^2(x)\, dx \int_a^b g^2(x)\, dx$$

for functions f and g, provided the integrals exist.

Other Forms of the Cauchy–Schwarz Inequality. We can obtain many particular forms of the Cauchy–Schwarz inequality by constructing different inner products. We begin with some inequalities for vectors in $\mathbb{R}^n$.

1. Take numbers $h_k > 0$. The inner product

$$\langle \mathbf{a}, \mathbf{b} \rangle = h_1 a_1 b_1 + \cdots + h_n a_n b_n$$

(the reader should verify satisfaction of the inner product axioms) yields

$$\left(\sum_{i=1}^{m} h_i a_i b_i \right)^2 \le \sum_{i=1}^{m} h_i a_i^2 \sum_{i=1}^{m} h_i b_i^2 \tag{3.26}$$

as another form of the Cauchy–Schwarz inequality.

2. Take a nondegenerate matrix $M = (m_{ij})_{i,j=1}^{n}$. Introducing the inner product

$$\langle \mathbf{a}, \mathbf{b} \rangle = \sum_{i=1}^{n} \left(\sum_{j=1}^{n} m_{ij} a_j \sum_{k=1}^{n} m_{ik} b_k \right)$$

(again, the reader should verify axioms P_1–P_3), we obtain

$$\left[\sum_{i=1}^{n}\left(\sum_{j=1}^{n}m_{ij}a_j\sum_{k=1}^{n}m_{ik}b_k\right)\right]^2 \le \sum_{i=1}^{n}\left(\sum_{j=1}^{n}m_{ij}a_j\right)^2\sum_{i=1}^{n}\left(\sum_{j=1}^{n}m_{ij}b_j\right)^2 \qquad (3.27)$$

as a form of the Cauchy–Schwarz inequality.

Note that equality holds in (3.26) and (3.27) if **a** and **b** are proportional or one of these vectors is zero. Proportionality means that there is a constant β such that $a_k = \beta b_k$ for all k.

We can immediately write down the series and integral versions of these two inequalities. The version of (3.26) for series is

$$\left(\sum_{i=1}^{\infty}h_ia_ib_i\right)^2 \le \sum_{i=1}^{\infty}h_ia_i^2\sum_{i=1}^{\infty}h_ib_i^2 . \qquad (3.28)$$

Assuming M is such that, for each n, the principal minor consisting of the elements of the first n rows and columns is nondegenerate, we obtain

$$\left[\sum_{i=1}^{\infty}\left(\sum_{j=1}^{\infty}m_{ij}a_j\sum_{k=1}^{\infty}m_{ik}b_k\right)\right]^2 \le \sum_{i=1}^{\infty}\left(\sum_{j=1}^{\infty}m_{ij}a_j\right)^2\sum_{i=1}^{\infty}\left(\sum_{j=1}^{\infty}m_{ij}b_j\right)^2 . \qquad (3.29)$$

In both inequalities, the series on the right should converge; as a consequence, we can state that the series on the left converge and the inequalities hold.

A limit passage applied to the corresponding Riemann sums gives

$$\left(\int_a^b h(x)f(x)g(x)\,dx\right)^2 \le \int_a^b h(x)f^2(x)\,dx\int_a^b h(x)g^2(x)\,dx , \qquad (3.30)$$

where we assume that $h(x) > 0$ over (a, b). The inequality makes sense when the integrals on the right are convergent. Note that it is possible to have $a = -\infty$ or $b = \infty$. We get a useful version of this inequality by putting $h(x) = x$ and considering the inequality over $(0, \infty)$:

$$\left(\int_0^{\infty} xf(x)g(x)\,dx\right)^2 \le \int_0^{\infty} xf^2(x)\,dx\int_0^{\infty} xg^2(x)\,dx . \qquad (3.31)$$

The integral form of (3.29) is left to the reader.

Application of the Cauchy–Schwarz Inequality to Matrices. On the set of $n \times n$ matrices with real elements, in a manner analogous to the case involving vectors, we can consider a number of inner products. Taking two real matrices $K = (k_{ij})_{i,j=1}^{n}$ and $M = (m_{ij})_{i,j=1}^{n}$ from the linear space of all such matrices, we can introduce an inner product

$$\langle M, N \rangle = \sum_{i,j=1}^{n} k_{ij}\,m_{ij} . \qquad (3.32)$$

It is easy to see that (3.32) does satisfy the inner product axioms. Hence we can immediately write down the Cauchy–Schwarz inequality for this case:

$$\left(\sum_{i,j=1}^{n} k_{ij}\, m_{ij}\right)^2 \le \sum_{i,j=1}^{n} k_{ij}^2 \sum_{i,j=1}^{n} m_{ij}^2 \,. \tag{3.33}$$

The difference between (3.33) and (3.25) is that we now have double summations. The infinite series version of (3.33) is

$$\left(\sum_{i,j=1}^{\infty} k_{ij}\, m_{ij}\right)^2 \le \sum_{i,j=1}^{\infty} k_{ij}^2 \sum_{i,j=1}^{\infty} m_{ij}^2 \tag{3.34}$$

and the integral version is

$$\left(\int_a^b \int_a^b f(x,y)g(x,y)\,dx\,dy\right)^2 \le \int_a^b \int_a^b f^2(x,y)\,dx\,dy \int_a^b \int_a^b g^2(x,y)\,dx\,dy \,. \tag{3.35}$$

In fact, (3.35) also holds on any domain $V \subseteq \mathbb{R}^n$ if the integrals make sense[1] and are convergent:

$$\left(\int_V f(\mathbf{x})g(\mathbf{x})\,dV\right)^2 \le \int_V f^2(\mathbf{x})\,dV \int_V g^2(\mathbf{x})\,dV \,.$$

To prove this, we introduce an inner product

$$\langle f_1, f_2 \rangle = \int_V f_1(\mathbf{x})f_2(\mathbf{x})\,dV$$

on the space of functions $f_k = f_k(\mathbf{x})$ that are square-integrable on V, and apply the Cauchy–Schwarz inequality (1.24).

We can introduce other forms for inner products over the sets of square and rectangular matrices. Weight factors (i.e., numerical factors with which the axioms P_1–P_3 hold) may be included as well. In this way, many inequalities for matrices (including infinite matrices) can be obtained.

3.8 Chebyshev's Inequality

Theorem 3.8 (Chebyshev's Inequality). *Let a_i and b_i be similarly ordered such that either*

$$\begin{cases} a_1 \le \cdots \le a_m \,, \\ b_1 \le \cdots \le b_m \,, \end{cases} \quad or \quad \begin{cases} a_1 \ge \cdots \ge a_m \,, \\ b_1 \ge \cdots \ge b_m \,. \end{cases}$$

---

[1] Here V should be sufficiently uncomplicated that we can introduce the integral.

Then

$$\frac{1}{m} \sum_{i=1}^{m} a_i b_i \geq \frac{1}{m} \sum_{i=1}^{m} a_i \cdot \frac{1}{m} \sum_{n=1}^{m} b_i$$

with equality if and only if $a_1 = \cdots = a_m$ or $b_1 = \cdots = b_m$.

Proof. For either of the two cases it is evident that for any choice of i, j,

$$(a_i - a_j)(b_i - b_j) \geq 0 .$$

Summation over both indices yields

$$\sum_{i=1}^{m} \sum_{j=1}^{m} (a_i - a_j)(b_i - b_j) \geq 0$$

and expansion gives

$$\sum_{i=1}^{m} a_i b_i \sum_{j=1}^{m} (1) - \sum_{i=1}^{m} a_i \sum_{j=1}^{m} b_j - \sum_{j=1}^{m} a_j \sum_{i=1}^{m} b_i + \sum_{j=1}^{m} a_j b_j \sum_{i=1}^{m} (1) \geq 0$$

or

$$2m \sum_{i=1}^{m} a_i b_i - 2 \sum_{i=1}^{m} a_i \sum_{i=1}^{m} b_i \geq 0 ,$$

as required.   □

Example 3.4. Choosing $b_i = a_i$ for all i, we get

$$\left(\frac{1}{m} \sum_{i=1}^{m} a_i \right)^2 \leq \frac{1}{m} \sum_{i=1}^{m} a_i^2 .$$

The square of an arithmetic mean cannot exceed the mean of the squares.   □

Example 3.5. Let us reconsider (1.41):

$$\tfrac{1}{3}(a^5 + b^5 + c^5) \geq \tfrac{1}{3}(a^2 + b^2 + c^2) \cdot \tfrac{1}{3}(a^3 + b^3 + c^3) \qquad (a, b, c > 0) .$$

Since this inequality is symmetric in its variables a, b, c, we can assume $a \leq b \leq c$. This implies both $a^2 \leq b^2 \leq c^2$ and $a^3 \leq b^3 \leq c^3$. Theorem 3.8 applies; equality holds if and only if $a = b = c$.   □

With functions $f(x)$ and $g(x)$, analogous operations yield

$$\int_a^b f(x)g(x)\,dx \geq \frac{1}{b-a} \int_a^b f(x)\,dx \int_a^b g(x)\,dx$$

if $f(x)$ and $g(x)$ are either both increasing or both decreasing on $[a, b]$. If one function is increasing and the other is decreasing, the inequality sign is reversed.

3.9 Jensen's Inequality

A function $f(x)$ is *convex* on the open interval (a, b) if and only if the inequality

$$f(px_1 + (1 - p)x_2) \le pf(x_1) + (1 - p)f(x_2) \qquad (3.36)$$

holds for all $x_1, x_2 \in (a, b)$ and every $p \in (0, 1)$. In the case of strict inequality for $x_1 \ne x_2$, f is *strictly convex* on (a, b). We note that any $x_p \in (x_1, x_2)$ can be expressed as $x_p = x_1 + (1 - p)(x_2 - x_1) = px_1 + (1 - p)x_2$ for some $p \in (0, 1)$. The straight line connecting the points $(x_1, f(x_1))$ and $(x_2, f(x_2))$ is

$$f_s(x) = f(x_1) + \left[\frac{f(x_2) - f(x_1)}{x_2 - x_1} \right] (x - x_1)$$

so that $f_s(x_p) = pf(x_1) + (1 - p)f(x_2)$. Geometrically, convexity prevents the graph of f from rising above the secant line connecting any two of its points (Fig. 3.2).

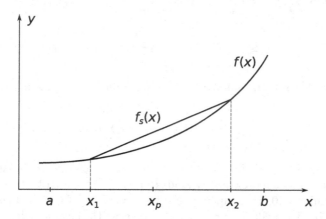

Fig. 3.2 Function convexity. The graph of $f(x)$ does not rise above the secant line $f_s(x)$

Upon reflection it seems natural to associate convexity with the requirement that $f''(x) \ge 0$ on (a, b). In fact, this is equivalent to (3.36) for functions twice continuously differentiable on (a, b) (Problem 3.26(c)). We also mention that other definitions of convexity exist. An example is the *midpoint convexity* definition requiring that

$$f\left(\frac{x_1 + x_2}{2} \right) \le \frac{f(x_1) + f(x_2)}{2}$$

for $x_1, x_2 \in (a, b)$. Here the geometric requirement is only that the midpoint of every secant line lie on or above the graph of f. For more detailed information on convexity, see Mitrinovic [56]. Our main result for convex functions is as follows:

Theorem 3.9 (Jensen's Inequality). *Let f be convex on (a, b), let $x_1, \ldots, x_m$ be m points of (a, b), and let $\delta_1, \ldots, \delta_m$ be nonnegative constants such that $\delta_1 + \cdots + \delta_m = 1$. Then*

$$f\left(\sum_{i=1}^{m} \delta_i x_i\right) \le \sum_{i=1}^{m} \delta_i f(x_i) \,. \tag{3.37}$$

If f is strictly convex and if additionally each $\delta_i > 0$, then equality holds if and only if $x_1 = \cdots = x_m$.

Proof. Note that the case $\delta_m = 1$ is trivial, as we would have all the rest of the $\delta_k = 0$. We first consider the case for which $m = 2$ and proceed by induction. Now (3.37) holds by the convexity of f. If $x_1 = x_2$, then equality holds in (3.37) by inspection, and if f is strictly convex, all $\delta_i > 0$, and equality holds in (3.37) we must have $x_1 = x_2$, for otherwise

$$f(\delta_1 x_1 + \delta_2 x_2) < \delta_1 f(x_1) + \delta_2 f(x_2) \,.$$

Now assume the theorem is true for $m = k$ and suppose $\delta_1 + \cdots + \delta_{k+1} = 1$. We have

$$f\left(\sum_{i=1}^{k+1} \delta_i x_i\right) = f\left((1 - \delta_{k+1}) \sum_{i=1}^{k} \frac{\delta_i}{1 - \delta_{k+1}} x_i + \delta_{k+1} x_{k+1}\right)$$

$$\le (1 - \delta_{k+1}) f\left(\sum_{i=1}^{k} \frac{\delta_i}{1 - \delta_{k+1}} x_i\right) + \delta_{k+1} f(x_{k+1})$$

by convexity of f. Since the numbers $\delta_i/(1 - \delta_{k+1})$ for $1 \le i \le k$ sum to 1,

$$f\left(\sum_{i=1}^{k} \frac{\delta_i}{1 - \delta_{k+1}} x_i\right) \le \frac{1}{1 - \delta_{k+1}} \sum_{i=1}^{k} \delta_i f(x_i) \,, \tag{3.38}$$

hence (with $m = k + 1$) (3.37) holds. If $x_1 = \cdots = x_{k+1}$, then equality holds in (3.37) by inspection. Now suppose equality holds (with $m = k + 1$) in (3.37), f is strictly convex, and all $\delta_i > 0$. Then equality also holds in (3.38), for if not then equality cannot hold in (3.37) either, contrary to hypothesis. Hence, since the theorem is assumed true for k numbers, $x_1 = \cdots = x_k$. Putting this into (3.37), we obtain

$$f\left(\left(\sum_{i=1}^{k} \delta_i\right) x_1 + \delta_{k+1} x_{k+1}\right) = \left(\sum_{i=1}^{k} \delta_i\right) f(x_1) + \delta_{k+1} f(x_{k+1}) \,,$$

so by the case $m = 2$, $x_{k+1} = x_1$ and hence the induction is complete. The other case, for which $\delta_m = 1$, is much easier; for then $\delta_1 = \cdots = \delta_{m-1} = 0$, whence (3.37) becomes simply $f(x_m) \le f(x_m)$.   □

Example 3.6. For $n \in \mathbb{N}$ we have

$$\left(\frac{x + y}{2}\right)^n \le \frac{x^n + y^n}{2} \qquad (x, y > 0)$$

because $f(x) = x^n$ is convex on $(0, \infty)$.   □

Example 3.7. Choosing $\delta_i = 1/m$ for $i = 1, \ldots, m$, we have

$$f\left(\frac{1}{m}\sum_{i=1}^{m} x_i\right) \le \frac{1}{m}\sum_{i=1}^{m} f(x_i)$$

for any convex f. If instead f is "concave" such that $-f$ is convex, our inequality becomes

$$f\left(\frac{1}{m}\sum_{i=1}^{m} x_i\right) \ge \frac{1}{m}\sum_{i=1}^{m} f(x_i) \ .$$

An example is $f(x) = \sin x$ on $(0, \pi)$, and we have

$$\frac{1}{m}\sum_{i=1}^{m} \sin\theta_i \le \sin\left(\frac{1}{m}\sum_{n=1}^{m}\theta_i\right) \qquad (0 < \theta_1 \le \cdots \le \theta_m < \pi) \ . \qquad (3.39)$$

In fact, the function $-\sin x$ is strictly convex on $(0, \pi)$ and equality holds in (3.39) only if $\theta_1 = \cdots = \theta_m$. See Problem 3.25 for an application. $\qquad\square$

An integral form of Jensen's inequality is introduced in Problem 3.28.

3.10 Friedrichs- and Poincaré-Type Inequalities

Suppose f is integrable on an interval (a, b). In this section we present two integral inequalities that provide estimates of the mean-square value of f in terms of integrals over its squared derivative f'^2. These results, along with their extensions (to higher dimensions, as in (6.8) and (6.12), as well as to other results such as Korn's inequality (6.18)), are widely used in continuum mechanics (see, e.g., [47, 77]).

Theorem 3.10 can be regarded as a simple case of *Friedrichs' inequality*.

Theorem 3.10. *If $f(a) = 0$, then*

$$\int_a^b f^2(x)\,dx \le \frac{(b-a)^2}{2}\int_a^b f'^2(x)\,dx \ . \qquad (3.40)$$

Proof. For $f(a) = 0$ and $x \ge a$, the Newton–Leibniz identity gives

$$f(x) = \int_a^x f'(t)\,dt \ .$$

We square both sides and apply the Schwarz inequality to get

$$f^2(x) = \left(\int_a^x 1 \cdot f'(t)\,dt\right)^2 \le \int_a^x 1^2 dt \int_a^x f'^2(t)\,dt \le (x-a)\int_a^b f'^2(t)\,dt \ ,$$

since $f'^2(t) \geq 0$. Finally, we integrate over x from a to b and obtain

$$\int_a^b f^2(x)\, dx \leq \int_a^b f'^2(t)\, dt \int_a^b (x - a)\, dx = \frac{(b - a)^2}{2} \int_a^b f'^2(t)\, dt\,,$$

which is (3.40).   □

Now suppose f is not restricted at the point a. If we take $f_c(x) = c = \text{constant}$, we see that for f_c the inequality (3.40) is incorrect. The correct form of a similar inequality for nonrestricted functions is the following one-dimensional version of *Poincaré's inequality*.

Theorem 3.11. *We have*

$$\int_a^b f^2(x)\, dx \leq \frac{1}{b - a} \left(\int_a^b f(x)\, dx \right)^2 + \frac{(b - a)^2}{2} \int_a^b f'^2(x)\, dx\,. \tag{3.41}$$

Proof. Squaring both sides of the identity

$$f(x_2) - f(x_1) = \int_{x_1}^{x_2} 1 \cdot f'(x)\, dx \qquad (x_1 \leq x_2)$$

and using the Schwarz inequality, we get

$$f^2(x_2) + f^2(x_1) - 2f(x_1)f(x_2) \leq \int_{x_1}^{x_2} 1^2 dx \int_{x_1}^{x_2} f'^2(x)\, dx$$

$$\leq (b - a) \int_a^b f'^2(x)\, dx\,.$$

Integrating with respect to x_2 over $[a, b]$, we obtain

$$\int_a^b f^2(x_2)\, dx_2 + (b - a)f^2(x_1) - 2f(x_1) \int_a^b f(x_2)\, dx_2$$

$$\leq (b - a)^2 \int_a^b f'^2(x)\, dx\,.$$

Finally, integrating with respect to x_1 over $[a, b]$, we have

$$(b - a) \int_a^b f^2(x_2)\, dx_2 + (b - a) \int_a^b f^2(x_1)\, dx_1 - 2 \int_a^b f(x_1)\, dx_1 \int_a^b f(x_2)\, dx_2$$

$$\leq (b - a)^3 \int_a^b f'^2(x)\, dx\,.$$

Rearrangement gives (3.41).   □

We will state a two-dimensional version of Poincaré's inequality on p. 169.

3.11 Problems

3.1. The following problems are related to Young's inequality.

(a)   Verify that $f(x) = x^{p-1}$ and $g(x) = x^{q-1}$ are mutually inverse if p and q are conjugate exponents.

(b)   Show that for any $\varepsilon > 0$,

$$ab \le \frac{\varepsilon a^p}{p} + \frac{\varepsilon^{1-q} b^q}{q}$$

where $a, b \ge 0$ and p and q are conjugate exponents.

(c)   Show that if $x, y \ge 0$ then $xy \le (x+1)\ln(x+1) - x + e^y - y - 1$.

(d)   Use the concavity of the log function to derive (3.13).

3.2. Assume $a, b, c, d > 0$ and prove the following:

(a) $a^4 + b^4 + c^4 + d^4 \ge 4abcd$,

(b) $(a+b)(b+c)(c+a) \ge 8abc$,

(c) $(ab + bc + ca)(a^4 + b^4 + c^4) \ge 9a^2 b^2 c^2$,

(d) $a + 1/a + b + 1/b \ge 2\sqrt{ab} + 2/\sqrt{ab}$,

(e) $(a^2 - b^2)^2 \ge 4ab(a-b)^2$,

(f) $\sqrt{(a+c)(b+d)} \ge \sqrt{ab} + \sqrt{cd}$ with equality if and only if $bc = ad$,

(g) $a/\sqrt{b} + b/\sqrt{a} \ge 2(ab)^{1/4}$,

(h) $(ab + cd)(ac + bd) \ge 4abcd$,

(i) $(a^2 b + b^2 c + c^2 a)(ab^2 + bc^2 + ca^2) \ge 9a^2 b^2 c^2$,

(j) $bc/a + ac/b + ab/c \ge a + b + c$.

3.3. Use AM–GM to prove the following inequalities [43, 56, 83].

(a)   For any natural number $n > 1$,

$$n! < \left(\frac{n+1}{2}\right)^n .$$

Also

$$(2n-1)!! < n^n \quad \text{and} \quad (n+1)^n > (2n)!!$$

where

$$(2n-1)!! = (2n-1) \cdot (2n-3) \cdot (2n-5) \cdots 5 \cdot 3 \cdot 1 ,$$
$$(2n)!! = (2n) \cdot (2n-2) \cdot (2n-4) \cdots 4 \cdot 2 .$$

(b)   If $a, b \ge 0$ and $n > 1$ is a natural number, then

$$\frac{a + (n-1)b}{n} \ge \left(ab^{n-1}\right)^{1/n}$$

with equality if and only if $a = b$.

(c)   For any natural number $n > 1$ we have

$$n^n \left[\tfrac{1}{2}(n+1)\right]^{2n} > (n!)^3 .$$

(d)   For any natural number $n > 1$ we have

$$n! > (n+1)^{\frac{1}{2}(n-1)} .$$

(e)   (*Korovkin's inequality*, [13, 43].) If $x_1, \ldots, x_n > 0$, then

$$\frac{x_1}{x_2} + \frac{x_2}{x_3} + \cdots + \frac{x_{n-1}}{x_n} + \frac{x_n}{x_1} \geq n$$

with equality if and only if $x_1 = \cdots = x_n$.

(f)   If $x_i > 0$ for $i = 1, \ldots, n$, then

$$n x_1 x_2 \cdots x_n \leq x_1^n + x_2^n + \cdots + x_n^n .$$

(g)   If $a > 0$, then

$$n a^{n+1} + 1 \geq a^n (n+1) .$$

(h)   If $a_i > 0$ for $i = 1, 2, \ldots, n$ and $n = 3, 4, \ldots$, then

$$n \sqrt[n]{a_1 \cdots a_n} - (n-1) \sqrt[n-1]{a_1 \cdots a_{n-1}} \leq a_n .$$

(i)   If the product of N positive numbers equals 1, then the sum of those numbers cannot be less than N.

(j)   The sequence $(1 - n^{-1})^n$ is monotone increasing.

(k)   If $a_1, \ldots, a_N$ are positive numbers that sum to 1 and m is a positive integer, then

$$\sum_{n=1}^{N} a_n^{-m} \geq N^{m+1}.$$

(l)   For any natural number n, we have

$$n! \leq 2(n/2)^n .$$

Remark: For large n, an asymptotic expression for $n!$ is given by *Stirling's formula*

$$n! \sim \sqrt{2\pi n}\,(n/e)^n , \quad \text{which means that } \lim_{n \to \infty} \frac{n!}{\sqrt{2\pi n}\,(n/e)^n} = 1 .$$

3.4. The following are simple applications of the AM–GM inequality.

(a)   Show that of all rectangles having a given area, a square has the least perimeter.

(b)   A charge q is removed from a given electric charge Q to make two separate charges q and $Q - q$. Determine q so that repulsion between the charges at a given distance is maximized.

3.5. Prove (3.7)

(a)   by induction, and

(b)   by the Lagrange multiplier technique.

3.6. Prove the weighted AM–GM inequality by induction on n.

3.7. Let $n \in \mathbb{N}$, and let $x_1, \ldots, x_n$ and $\delta_1, \ldots, \delta_n$ be positive numbers such that $\sum_{i=1}^n \delta_i = 1$. For any real number $t \neq 0$ define

$$g(t) = \left(\sum_{i=1}^{n} \delta_i x_i^t \right)^{1/t} .$$

(a)   Show that

$$g(t) \to \prod_{i=1}^{n} x_i^{\delta_i} \text{ as } t \to 0 \text{ so that we may define } g(0) = \prod_{i=1}^{n} x_i^{\delta_i}.$$

(b)   Show that g is increasing. Preliminary hint: Take the logarithm and use l'Hôpital's monotone rule on $(0, \infty)$ and $(-\infty, 0)$.

(c)   Note that $g(-1) \le g(0) \le g(1)$ gives

$$\left(\sum_{i=1}^{n} \delta_i / x_i \right)^{-1} \le \prod_{i=1}^{n} x_i^{\delta_i} \le \sum_{i=1}^{n} \delta_i x_i,$$

which is the weighted harmonic–geometric–arithmetic means inequality.

3.8. Let f be continuous on $[a, b]$ with $f(x) > 0$ on $[a, b]$. Prove

$$\frac{b-a}{\int_a^b (1/f(x))\,dx} \le \exp\left[\frac{1}{b-a} \int_a^b \ln f(x)\,dx \right] \le \frac{1}{b-a} \int_a^b f(x)\,dx.$$

This is the harmonic–geometric–arithmetic mean inequality for integrals.

3.9. Let $n \in \mathbb{N}$, $n \ge 2$, and $x_1, \ldots, x_n$ be positive numbers. On $(0, \infty)$ define

$$h(t) = \left(\sum_{i=1}^{n} x_i \right)^{1/t}.$$

Show that h is decreasing.

3.10. Use (3.8) to prove the following [36, 43, 81].

(a)   If $a_1, \ldots, a_n$ are positive numbers whose sum is s, then

$$\frac{1}{a_1} + \frac{1}{a_2} + \cdots + \frac{1}{a_n} \ge \frac{n^2}{s}.$$

(b)   If $a, b \ge 0$ and $a + b \ge 1$, then $a^4 + b^4 \ge 1/8$.

(c)   If $a, b, c > 0$, then

$$\sqrt{\frac{a}{b+c}} + \sqrt{\frac{b}{c+a}} + \sqrt{\frac{c}{a+b}} > 2.$$

(d)   If $x_1, \ldots, x_n$ are positive numbers that sum to 1, then

$$\left(x_1 + \frac{1}{x_1} \right)^2 + \cdots + \left(x_n + \frac{1}{x_n} \right)^2 \ge \frac{(n^2 + 1)^2}{n}.$$

3.11. Show that for any m numbers a_i that satisfy $0 < a_1 < \cdots < a_m$ and any m positive numbers λ_i that sum to 1, *Kantorovich's inequality*

$$\sum_{i=1}^{m} \lambda_i a_i \sum_{i=1}^{m} \frac{\lambda_i}{a_i} \le \left(\frac{A}{G} \right)^2$$

holds, where $A = (a_1 + a_m)/2$ and $G = \sqrt{a_1 a_m}$.

3.12. Suppose p and q are positive real numbers such that $p^{-1} + q^{-1} = 1$. Let $a_1, \ldots, a_m$ be nonzero numbers. Define $b_i = c|a_i|^{p-1}$ for $i = 1, \ldots, m$. Verify that

$$\frac{|a_i|^p}{\sum_{i=1}^{m} |a_i|^p} = \frac{|b_i|^p}{\sum_{i=1}^{m} |b_i|^p} \qquad (i = 1, \ldots, m) . \tag{3.42}$$

By Theorem 3.5 equality must hold in Hölder's inequality. Verify this by direct substitution. Conversely, show that (3.42) implies there is a $c > 0$ such that

$$|b_i| = c|a_i|^{p-1} \tag{3.43}$$

for $i = 1, \ldots, m$. Hence the condition for equality in Hölder's inequality can be stated by (3.43).

3.13. Use Lagrange multipliers to verify Hölder's inequality.

3.14. Justify Eq. (3.14).

3.15. Use Hölder's inequality to show that

$$\left(\sum_{k=1}^{n} |a_k| \right)^p \leq n^{p-1} \sum_{k=1}^{n} |a_k|^p \qquad (p \geq 1) .$$

3.16. Given n real m-tuples of positive numbers,

$$\{a_1^{(k)}, \ldots, a_m^{(k)}\} \qquad (k = 1, \ldots, n) ,$$

show that Minkowski's inequality for sums can be generalized as

$$\left[\sum_{i=1}^{m} \left(\sum_{k=1}^{n} a_i^{(k)} \right)^p \right]^{1/p} \leq \sum_{k=1}^{n} \left(\sum_{i=1}^{m} (a_i^{(k)})^p \right)^{1/p} \qquad (p > 1) . \tag{3.44}$$

3.17. Show that the Cauchy–Schwarz inequality (3.25) follows from the *Lagrange identity*

$$\sum_{i=1}^{n} a_i^2 \sum_{i=1}^{n} b_i^2 - \left(\sum_{i=1}^{n} a_i b_i \right)^2 = \sum_{\substack{i,j=1 \\ i<j}}^{n} (a_i b_j - a_j b_i)^2 .$$

3.18. A function f is *square integrable* on $[a, b]$ if

$$\int_a^b |f(x)|^2 \, dx < \infty .$$

Show that the sum of two square integrable functions is square integrable.

3.19. Prove that if $h(x) \geq 0$, then

$$\left| \int_a^b f(x) g(x) h(x) \, dx \right|^2 \leq \int_a^b f^2(x) h(x) \, dx \int_a^b g^2(x) h(x) \, dx .$$

3.20. Use the Cauchy–Schwarz inequality to prove the following statements. Assume all variables are positive.

(a)

$$n^2 \leq \sum_{i=1}^{n} c_i \sum_{i=1}^{n} \frac{1}{c_i} ,$$

(b)

$$\sqrt{(a+b)(c+d)} \geq \sqrt{ac} + \sqrt{bd} ,$$

(c)

$$a\sqrt{a^2+c^2} + b\sqrt{b^2+c^2} \leq a^2 + b^2 + c^2 ,$$

(d)

$$\frac{1}{\sqrt{bc}} + \frac{1}{\sqrt{ca}} + \frac{1}{\sqrt{ab}} \leq \frac{1}{a} + \frac{1}{b} + \frac{1}{c} ,$$

(e)   If $a > c$ and $b > c$, then

$$\sqrt{c(a-c)} + \sqrt{c(b-c)} \leq \sqrt{ab} .$$

3.21. [57] Obtain the following as consequences of the Cauchy–Schwarz inequality, assuming a_k, b_k, c_k, d_k are positive real numbers.

(a)

$$\left(\sum_{k=1}^{n} a_k b_k\right)^2 \leq \sum_{k=1}^{n} k a_k^2 \sum_{k=1}^{n} \frac{b_k^2}{k} ,$$

(b)

$$\left(\sum_{k=1}^{n} a_k^m\right)^2 \leq \sum_{k=1}^{n} a_k^{m+s} \sum_{k=1}^{n} a_k^{m-s} ,$$

(c)

$$\left(\sum_{k=1}^{n} \frac{a_k}{k}\right)^2 \leq \sum_{k=1}^{n} k^3 a_k^2 \sum_{k=1}^{n} \frac{1}{k^5} ,$$

(d)

$$\left(\sum_{k=1}^{n} a_k b_k c_k\right)^4 \leq \sum_{k=1}^{n} a_k^4 \sum_{k=1}^{n} b_k^4 \left(\sum_{k=1}^{n} c_k^2\right)^2 ,$$

(e)

$$\left(\sum_{k=1}^{n} a_k\right)^2 \leq n \sum_{k=1}^{n} a_k^2 ,$$

(f)

$$\sum_{k=1}^{n} \sqrt{a_k b_k} \leq \sqrt{\sum_{k=1}^{n} a_k} \sqrt{\sum_{k=1}^{n} b_k} ,$$

(g)

$$\left(\sum_{k=1}^{n} a_k b_k c_k d_k\right)^4 \leq \sum_{k=1}^{n} a_k^4 \sum_{k=1}^{n} b_k^4 \sum_{k=1}^{n} c_k^4 \sum_{k=1}^{n} d_k^4 ,$$

(h)

$$\left(\sum_{k=1}^{n} a_k b_k\right)^2 \leq \sum_{k=1}^{n} a_k \sum_{k=1}^{n} a_k b_k^2 ,$$

(i)

$$\sum_{k=1}^{n} \frac{b_k^2}{a_k} \geq \frac{(\sum_{k=1}^{n} b_k)^2}{\sum_{k=1}^{n} a_k} .$$

3.22. Show that if g is positive on $[a, b]$ and $\int_a^b g(x)\,dx = 1$, then

$$\int_a^b f(x)\,dx \leq \left(\int_a^b \frac{f^2(x)}{g(x)}\,dx\right)^{1/2} .$$

3.23. [27] Show that if $r > 1$ then $2(r-1) \leq (r+1)\ln r$.

3.24. Use the Chebyshev inequalities for integrals to derive the inequalities

$$\int_a^b [f(x)]^2 \, dx \ge \frac{1}{b-a} \left(\int_a^b f(x) \, dx \right)^2$$

and

$$\int_a^b f(x) \, dx \int_a^b \frac{dx}{f(x)} \ge (b-a)^2.$$

3.25. Prove that of all N sided polygons that can be inscribed in a circle of fixed radius, a regular polygon has the greatest area.

3.26. Establish the following facts about convex functions.

(a)   Let $\alpha, \beta \ge 0$. If f, g are convex functions on (a, b), then so is $\alpha f + \beta g$.

(b)   If f_n is convex for each $n = 1, 2, \ldots$ and $\lim_{n \to \infty} f_n(x) = f(x)$ (pointwise convergence on (a, b)), then f is convex.

(c)   The condition

$$(1 - p)(x_2 - x_1)^2 \int_0^1 \int_{(1-p)s}^s f''(x_1 + (x_2 - x_1)t) \, dt \, ds \ge 0$$

is equivalent to (3.36) for functions f that are twice continuously differentiable. Hence, $f''(x) \ge 0$ is necessary and sufficient for the convexity of such functions.

(d)   The function $x \ln x$ is convex on $(0, \infty)$.

3.27. The following are applications of Jensen's inequality.

(a)   Show that for positive real numbers x_k we have

$$\left(\sum_{k=1}^n x_k \right)^2 \le n \sum_{k=1}^n x_k^2 \, .$$

(b)   Show that $x \ln x + y \ln y \ge (x + y) \ln[(x + y)/2]$ for $x, y > 0$.

(c)   Use Jensen's inequality with $f(x) = -\ln x \; (x > 0)$ to deduce (3.4).

(d)   [67] Show that

$$1 + \exp\left(\frac{1}{n} \sum_{k=1}^n x_k \right) \le \left\{ \prod_{k=1}^n [1 + \exp(x_k)] \right\}^{1/n} \, .$$

3.28. Let g and p be continuous and defined for $a \le t \le b$ such that $\alpha \le g(t) \le \beta$ and $p(t) > 0$. Let f be continuous and convex on the interval $\alpha \le u \le \beta$. Show that

$$f \left(\frac{\int_a^b g(t) p(t) \, dt}{\int_a^b p(t) \, dt} \right) \le \frac{\int_a^b f(g(t)) \, p(t) \, dt}{\int_a^b p(t) \, dt} \, .$$

This is Jensen's inequality for integrals.

Chapter 4
Inequalities in Abstract Spaces

4.1 Introduction

Generality is gained by working in abstract spaces. For instance, all essential aspects of the topics of convergence and continuity can be studied in the context of a metric space. When we search for solutions to problems of physical interest, we must often search among the members of linear spaces (also known as vector spaces). Inequalities provide basic structure for abstract spaces like these, and we turn to a consideration of that topic in the present chapter. In doing so we present a few topics from functional analysis. Needless to say our coverage is neither broad nor deep: we hope only to catch a glimpse of inequalities in the kind of abstract setting that can unify many of our previous results before we proceed to the chapter on applications.

We begin by briefly comparing certain aspects of finite and infinite dimensional spaces, although we have implicitly used the latter when considering inequalities for sequences and integrals.

4.2 Vectors and Norms

We deduced some inequalities for finite numbers of variables, regarding the elements of an ordered set $(a_1, \ldots, a_n)$ as the components of a vector in some basis of $\mathbb{R}^n$ or $\mathbb{C}^n$. While it is natural at such times to assume the basis is orthonormal, the idea of working with vectors—not merely with their components relative to some fixed basis—leads us immediately to inequalities that may not formally resemble those we started with.

Example 4.1. If we expand $\mathbf{a}$ and $\mathbf{b}$ in a nonorthogonal basis $(\mathbf{e}_1, \ldots, \mathbf{e}_n)$ as

$$\mathbf{a} = \sum_{k=1}^{n} a_k \mathbf{e}_k, \qquad \mathbf{b} = \sum_{k=1}^{n} b_k \mathbf{e}_k,$$

M.J. Cloud et al., *Inequalities: With Applications to Engineering*,
DOI 10.1007/978-3-319-05311-0_4, © Springer International Publishing AG 2014

the Cauchy–Schwarz inequality $|\mathbf{a} \cdot \mathbf{b}| \leq \|\mathbf{a}\| \|\mathbf{b}\|$ takes the form

$$\left| \sum_{k,m=1}^{n} g_{km} a_k b_m \right| \leq \left(\sum_{k,m=1}^{n} g_{km} a_k a_m \right)^{1/2} \left(\sum_{k,m=1}^{n} g_{km} b_k b_m \right)^{1/2}$$

where the quantities $g_{km} = \mathbf{e}_k \cdot \mathbf{e}_m$ are called the metric coefficients.   □

So we wish to discuss the vector concept. In elementary mathematics, students start with the idea that vectors are arrows in a plane or in space, carrying attributes of length and the possibility of being multiplied by scalars and added together using the parallelogram rule. It may be emphasized that the rules for working with vectors were borrowed from those for forces considered as vectorial quantities. Later, in a course on linear algebra, students are exposed to the set of axioms that are obeyed by vectors when they constitute a linear space. We assume the reader is aware of these axioms. In fact, the vectors traditionally used in applications differ somewhat from the abstract vectors of linear algebra. First, they are dimensional quantities carrying physical units. Second, they may possess additional properties by virtue of their attachment to physical objects; if a force acting on a rigid body is shifted parallel to itself, a compensating moment must be introduced in order to produce an equivalent action on the body. But the most important property of any vector (physical or abstract) is its invariance under coordinate transformation. This invariance leads to the development of formidable looking transformation formulas when the theory of vectors is applied to the theory of functions taking vector values on some domain in $\mathbb{R}^n$ or $\mathbb{C}^n$. The reader is aware that any n-dimensional linear space X can be identified with $\mathbb{R}^n$ or $\mathbb{C}^n$, even if the elements of X differ in nature from their counterparts in $\mathbb{R}^n$ or $\mathbb{C}^n$. For example, the set of real polynomials $a_0 x^n + \cdots + a_n$ can be placed in one-to-one correspondence with the set of "vectors" $(a_0, \ldots, a_n)$ from $\mathbb{R}^{n+1}$, assuming a Cartesian basis for $\mathbb{R}^{n+1}$.

We said that in $\mathbb{R}^n$ we can introduce various norms fulfilling axioms N_1–N_3. When it is necessary to distinguish between the spaces based on $\mathbb{R}^n$ but with different norms, we use the notation $(\mathbb{R}^n, \|\cdot\|)$ to specify a particular space. We also said that all norms on $\mathbb{R}^n$ are equivalent (p. 22). For $\mathbf{x} = (x_1, \ldots, x_n) \in \mathbb{R}^n$, one commonly used norm is the p-norm

$$\|\mathbf{x}\|_p = \left(\sum_{k=1}^{n} |x_k|^p \right)^{1/p} \qquad (p \geq 1) .$$

Verification of axioms N_1 and N_2 is trivial, while N_3 amounts to Minkowski's inequality. These norms are typically used with canonical bases, but not always. It is worth noting that the p-norm expression is still meaningful when $p \to \infty$; in this case the value of the norm is given by

$$\|\mathbf{x}\|_\infty = \sup_{1 \leq k \leq n} |x_k| .$$

The norm concept is used to introduce the limit of a sequence of vectors $\{\mathbf{x}_i\}$, in complete analogy with the limit of a numerical sequence. That is, $\mathbf{x}^*$ is the limit of

{$\mathbf{x}_i$} if for any $\varepsilon > 0$ there is an integer N, dependent on ε, such that $\|\mathbf{x}^* - \mathbf{x}_i\| < \varepsilon$ whenever $i > N$. The equivalence of all norms on $\mathbb{R}^n$ means that a vector sequence is convergent relative to a given norm if and only if it is convergent relative to any other norm. Using the norm $\|\cdot\|_\infty$, we find that convergence of a sequence of vectors {$\mathbf{x}_i$} is equivalent to component-wise convergence (i.e., the convergence of each of the numerical component sequences). This provides some familiar results from elementary calculus, such as the fact that every Cauchy sequence of vectors in $\mathbb{R}^n$ has a limit in $\mathbb{R}^n$. (The reader can formulate the definition of a Cauchy sequence in $\mathbb{R}^n$ by adapting the ordinary definition of a numerical Cauchy sequence. It is merely necessary to replace the numerical sequence with a vector sequence and the absolute value function with the norm.) Another such fact is that if {$\mathbf{x}_i$} is bounded, it contains a Cauchy subsequence. In this way, we can reconsider all the facts from the calculus of multivariable functions.

Infinite-Dimensional Normed Sequence Spaces. Since Minkowski's inequality holds for sequences, we could consider a normed space of infinite-dimensional vectors by introducing the norm

$$\|\mathbf{x}\|_p = \left(\sum_{k=1}^{\infty} |x_k|^p \right)^{1/p} . \tag{4.1}$$

We would encounter difficulties, however. For example, the sequence $\mathbf{x}_0 = \{1/i\}$ has a finite norm when $p > 1$, but the norm $\|\mathbf{x}_0\|_1$ does not exist as the corresponding series diverges.

For this reason, when defining a normed sequence space we must restrict attention to a particular subset of infinite sequences. For $p \geq 1$, we define ℓ_p as the linear space of sequences for which

$$\sum_{k=1}^{\infty} |x_k|^p < +\infty ;$$

this ensures that $\|\mathbf{x}\|_p$ in (4.1) is finite for any $\mathbf{x} \in \ell_p$. Unlike the situation with $\mathbb{R}^n$, for different values of p the spaces ℓ_p do not coincide as sets. Normally, infinite sequences are not considered as infinite vectors (although we often use the vector notation and the term "vector") because the change of basis presents a problem. It may even present an impossibility, as is the case with the space m consisting of bounded sequences with the norm

$$\|\mathbf{x}\|_m = \sup_k |x_k| .$$

It may seem natural to introduce something like a canonical basis of infinite vectors $\mathbf{i}_k$, where $\mathbf{i}_k$ has 1 as its kth component and 0 as each of its remaining components. For a vector $\mathbf{x} \in \mathbb{R}^n$ we have, by definition, $\mathbf{x} = \sum_{k=1}^{n} x_k \mathbf{i}_k$. However, we cannot let $n \to \infty$ as the sum does not converge in the ordinary sense. Therefore $(\mathbf{i}_1, \mathbf{i}_2, \ldots)$ cannot be an ordinary basis in m.

The reader should be aware that in an infinite-dimensional normed space, some of the main theorems from elementary calculus are not valid. Most importantly, some normed spaces contain Cauchy sequences that fail to converge. If every Cauchy sequence in a normed space has a limit (in the space), the space is said to be *complete* or a *Banach space*. The spaces ℓ_p with $p \geq 1$, discussed above, are Banach spaces. According to Weierstrass' theorem, the space of functions f that are continuous on $[a, b]$ with the norm

$$\|f\| = \max_{x \in [a,b]} |f(x)|$$

is also a Banach space. Many useful function spaces are Banach spaces, as the reader can verify by consulting textbooks on functional analysis (e.g., [47, 48]).

Also important is the fact that in any infinite-dimensional normed space there are sequences that do not contain Cauchy subsequences. In such a space, the notion of a compact set cannot be equated with the notion of a closed and bounded set as is done in calculus.

Norm of a Matrix. Now we consider an $n \times n$ matrix A. We know that the product of $A = (a_{ij})_{i,j=1}^{n}$ and a column vector $\mathbf{x} = (x_1, \ldots, x_n)^T$ is a column vector

$$A\mathbf{x} = \left(\sum_{j=1}^{n} a_{1j}x_j , \; \ldots \; , \; \sum_{j=1}^{n} a_{nj}x_j \right)^T .$$

Here we use some fixed basis in $\mathbb{R}^n$ or $\mathbb{C}^n$; we will not consider the question of when the result $A\mathbf{x}$ does not depend on the basis, which leads to the notion of an operator in $\mathbb{R}^n$ or even to the broader notion of a tensor. It is clear that the set of $n \times n$ matrices, taken together with the actions of addition and multiplication by a scalar, form a linear space of dimension $n \times n$. On a finite dimensional space, there are infinitely many equivalent norms. Among these are norms that are compatible with the norm of a column vector $\mathbf{x}$:

$$\|A\| = \sup_{\|\mathbf{x}\| \neq 0} \frac{\|A\mathbf{x}\|}{\|\mathbf{x}\|} .$$

Equivalently, we can state that $\|A\|$ is a number such that for all $\mathbf{x} \in \mathbb{R}^n$ we have

$$\|A\mathbf{x}\| \leq \|A\| \, \|\mathbf{x}\| , \tag{4.2}$$

and, for any $\varepsilon > 0$ there is $\mathbf{x}^*$ such that

$$\|A\mathbf{x}\| > (\|A\| - \varepsilon) \, \|\mathbf{x}\| .$$

In fact, we may consider different norms for $\mathbf{x}$ and $A\mathbf{x}$.

Example 4.2. The matrix norm depends on the norm introduced on the space of vectors $\mathbf{x}$. Suppose the norm for $\mathbf{x}$ and $A\mathbf{x}$ is $\|\cdot\|_\infty$. Then

$$\|A\| = \max_{1 \leq k \leq n} \sum_{m=1}^{n} |a_{km}| .$$

Indeed

$$\|A\mathbf{x}\|_\infty = \max_{1\le k\le n}\left|\sum_{m=1}^{n}a_{km}x_m\right| \le \max_{1\le m\le n}|x_m|\cdot\max_{1\le k\le n}\left|\sum_{m=1}^{n}a_{km}\right| .$$

So inequality (4.2) holds. It remains to show that there is a nonzero vector for which equality is attained in (4.2). To construct the vector, we take k^* for which

$$\sum_{m=1}^{n}|a_{k^*m}| = \|A\| .$$

Next we take $1 = |x_1| = \cdots = |x_n|$ and such that $a_{k^*m}x_m = |a_{k^*m}x_m|$ for all m. For this vector we have equality in (4.2).   □

It is worth noting that if the vector norm is

$$\|\mathbf{x}\|_2 = \left(\sum_{k=1}^{n}|x_k|^2\right)^{1/2} ,$$

then the matrix norm is $\|A\| = \sqrt{\lambda}$, where λ is the maximum eigenvalue of the product AA^T.

4.3 Metric Spaces

Sometimes we must work within sets of elements that do not constitute linear spaces. For this reason it is useful to introduce the notion of metric space, of which the normed space is a particular case.

Let M be a nonempty set of elements (often called points). Let $d(x, y)$ be a real-valued function defined for each $x, y \in M$ such that:

M_1. $d(x, y) = 0$ if and only if $x = y$.

M_2. $d(x, y) = d(y, x)$.

M_3. $d(x, y) \le d(x, z) + d(z, y)$ for every $x, y, z \in S$.

Then M taken together with d is a *metric space* (M, d), and d is a distance or *metric* on M. When the context makes the choice of metric clear, we may simply refer to M as a metric space. Property M_3 is an abstract version of the triangle inequality. Putting $y = x$ in M_3 and using the other two properties, we get $d(x, z) \ge 0$ so that distance as defined is never negative. The distance is also symmetric by M_2, and we see that properties M_1–M_3 do satisfy our primary expectations about the distance concept.

Example 4.3. $\mathbb{R}$ and $\mathbb{C}$ are metric spaces, with distances each defined by the usual absolute value metric

$$d(x, y) = |x - y| .$$

A closed interval $[a, b]$ in $\mathbb{R}$ is also a metric space with this metric. The spaces $\mathbb{R}^n$ and $\mathbb{C}^n$, consisting of n-tuples of elements of $\mathbb{R}$ and $\mathbb{C}$ respectively, are metric spaces under the Euclidean metrics

$$d(x, y) = \left(\sum_{k=1}^{n} (x_k - y_k)^2 \right)^{1/2}, \qquad d(z, w) = \left(\sum_{k=1}^{n} |z_k - w_k|^2 \right)^{1/2}.$$

Note that $d(x, 0)$ looks just like the Euclidean norm in $\mathbb{R}^n$. Later we will see that if a linear space carries a norm $\|\cdot\|$, we can introduce the metric $d(x, y) = \|x - y\|$.   □

Example 4.4. The set of all real-valued functions defined and continuous on $[a, b]$ forms a metric space with the max metric

$$d(f, g) = \max_{x \in [a,b]} |f(x) - g(x)|.$$

The space is denoted $C[a, b]$. Let us check for satisfaction of the metric properties. We have $d(f, g) = 0$ if and only if $|f(x) - g(x)| = 0$ for all $x \in [a, b]$, verifying M_1. Property M_2 is obviously satisfied. For M_3, we have

$$|f(x) - g(x)| = |f(x) - h(x) + h(x) - g(x)| \le |f(x) - h(x)| + |h(x) - g(x)|$$

so that

$$\max_{x \in [a,b]} |f(x) - g(x)| \le \max_{x \in [a,b]} |f(x) - h(x)| + \max_{x \in [a,b]} |h(x) - g(x)|,$$

as desired.   □

We now state some definitions important in the study of metric spaces. These are related to the notion of limit for abstract elements such as the elements of sequence or function spaces. The ideas extend and mimic those from ordinary calculus. For the most part, we must replace the absolute value functions in the definitions with distance functions made available through the metrics that have been introduced for such elements. Hence we should begin with the notion of a neighborhood.

An ε-neighborhood of x_0 in M is a set

$$N_\varepsilon(x_0) = \{x \in M : d(x, x_0) < \varepsilon\}.$$

This is a direct extension of the corresponding definition in $\mathbb{R}$ (recall Example 1.11). A set S in M is *open* if, given any $x_0 \in S$, there exists $\varepsilon > 0$ such that $N_\varepsilon(x_0)$ is contained in S. A set S in M is said to be *closed* if its complement in M is open. A point $z \in M$ is a *limit point* (or *accumulation point*) of a set S if every ε-neighborhood of z contains at least one point of S distinct from z. It can be shown that S is closed if and only if it contains all its limit points. A sequence of points $\{x_n\}$ converges to the limit x if, for every $\varepsilon > 0$, there exists N such that $d(x_n, x) < \varepsilon$ whenever $n > N$. A sequence $\{x_n\}$ is a Cauchy sequence if, for every $\varepsilon > 0$, there exists N such that for every pair of numbers m, n, the inequalities $m > N$ and $n > N$ together imply that $d(x_m, x_n) < \varepsilon$.

Example 4.5. In $\mathbb{R}^3$, an ε-neighborhood of point $\mathbf{x}_0$ is a ball of radius ε centered at $\mathbf{x}_0$ if we take the Euclidean distance as the metric. If in $\mathbb{R}^3$ we take the metric

$$d(\mathbf{x}, \mathbf{y}) = \max_k |x_k - y_k|$$

(the reader should verify that this is a metric), then an ε-neighborhood is a cube centered at $\mathbf{x}_0$ with side length 2ε. Now we consider a more complicated example. An ε-neighborhood of a function f in $C[a, b]$ consists of all continuous functions whose graphs fall inside the curved band (Fig. 4.1)

$$S = \{(x, y): f(x) - \varepsilon < y < f(x) + \varepsilon, \quad a \le x \le b\}.$$

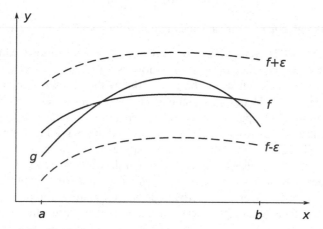

Fig. 4.1 An ε-neighborhood of a function f in $C[a, b]$. All continuous functions g whose graphs lie in the indicated band belong to the ε-neighborhood

On the set of absolutely integrable functions defined on $[a, b]$, we can introduce the metric

$$d(f, g) = \int_a^b |f(x) - g(x)| \, dx \tag{4.3}$$

(again, the reader should verify satisfaction of the metric axioms). Finally, an ε-neighborhood of a function f in this metric space consists of the set of all functions g satisfying the inequality

$$\int_a^b |g(x) - f(x)| \, dx < \varepsilon.$$

Here we cannot draw a picture for the ε-neighborhood of f, since for any (x, y), $x \in [a, b]$ we can find a function g that takes this value: $g(x) = y$. We are accustomed to think about small neighborhoods in calculus; here the smallness is in an integral sense but not in a local sense.

In metric spaces of functions, we can also think of the elements as "points," and even picture the convergence of a sequence of functions as we do the convergence of a sequence of real numbers. However, the reader should understand that this image represents a radical simplification of the actual picture. For example, when we consider the convergence of a sequence $\{f_n\}$ in $C[a, b]$, we find that for each $x \in [a, b]$ the numerical sequence $\{f_n(x)\}$ converges as $n \to \infty$. But with integral-type metrics such as (4.3), we can see convergence only in an "average" sense on $[a, b]$; there may be points x where $\{f_n(x)\}$ is not a numerical Cauchy sequence. □

Theorem 4.1. *If $\{x_n\}$ converges, then $\{x_n\}$ is a Cauchy sequence.*

Proof. Let $x_n \to x$. Then by the triangle inequality,

$$d(x_m, x_n) \leq d(x_m, x) + d(x, x_n) < \varepsilon/2 + \varepsilon/2 = \varepsilon$$

for sufficiently large n, m. This satisfies the definition of a Cauchy sequence.   □

The converse of Theorem 4.1 is false. However, in an application we may be able to prove that some sequence of approximations is a Cauchy sequence in a metric space, and in such a case it would be nice to know that the sequence has a limit. So it makes sense to select as a special class the spaces in which any Cauchy sequence has a limit. Such spaces are called *complete* metric spaces. The good news is that the spaces $\mathbb{R}^n$, $\mathbb{C}^n$, $C[a, b]$, and many others that are used in applications are complete. The bad news is that the functional spaces with integral-type metrics are incomplete if the Riemann integral and classical derivatives are used. As we said, complete spaces are much more convenient, and it has been shown that the Lebesgue integral and so-called generalized derivatives convert many incomplete spaces into complete spaces by extending the sets of functions that constitute the corresponding metric spaces.

Example 4.6. The space $M = C[a, b]$ is complete. Let $\{f_n\}$ be a Cauchy sequence in M. For each $x \in [a, b]$, $\{f_n(x)\}$ is a Cauchy sequence in $\mathbb{R}$ and hence has a limit which we denote by $\phi(x)$. To show that $\phi(x)$ is continuous at $x_1 \in [a, b]$ we use an $\varepsilon/3$ argument. For $x_2 \in [a, b]$

$$|\phi(x_1) - \phi(x_2)| \leq |\phi(x_1) - f_n(x_1)| + |f_n(x_1) - f_n(x_2)| + |f_n(x_2) - \phi(x_2)|.$$

The first and third terms can be made less than $\varepsilon/3$ for a sufficiently large choice of n, independent of x_1 and x_2, and with n fixed, the middle term can be made less than $\varepsilon/3$ by choosing x_2 sufficiently close to x_1. Also, if $K > 0$, then

$$\{f \in C[a, b] : |f(x)| \leq K \text{ for all } x \in [a, b]\}$$

is a complete metric space by the previous argument and Lemma 2.1.   □

Next, we take M_1, M_2 to be two metric spaces with distance functions d_1, d_2, respectively, and denote by $F: M_1 \to M_2$ a mapping (i.e., function) from M_1 to M_2. If $F: M \to M$, we say that F is a mapping on M. A mapping $F: M_1 \to M_2$ is continuous at $x_0 \in M_1$ if for every $\varepsilon > 0$, there is a $\delta > 0$ such that $d_2(F(x), F(x_0)) < \varepsilon$ whenever $d_1(x, x_0) < \delta$.

Lemma 4.1 (Persistence of Sign). *If M is a metric space and $f: M \to \mathbb{R}$ is continuous at x_0 with $f(x_0) > 0$, then there exists a $\delta > 0$ such that $f(x) > 0$ whenever $d(x, x_0) < \delta$.*

Proof. Analogous to the proof of Lemma 2.3. □

The relationship between convergence and continuity, noted in Theorem 2.1, extends to a general metric space.

Theorem 4.2. *The mapping $F: M_1 \to M_2$ is continuous at $x_0 \in M_1$ if and only if $F(x_n) \to F(x_0)$ whenever $x_n \to x_0$.*

Proof. Analogous to the proof of Theorem 2.1. ▪ □

Iteration in a Metric Space

Let F be a mapping on M. We say that F is a *contraction mapping* on M if there exists a number $\alpha \in [0, 1)$ such that

$$d(F(x_1), F(x_2)) \le \alpha d(x_1, x_2) \qquad (4.4)$$

whenever $x_1, x_2 \in M$. A point y is a *fixed point* of F if $F(y) = y$.

An iterative process is a method of locating a fixed point of F, or, in other words, of solving the equation $y = F(y)$. The method is based on the use of the recursion

$$y_{n+1} = F(y_n) \qquad (n = 0, 1, 2, \ldots)$$

to obtain successive approximations to a fixed point y. The construction of such a sequence is called *Picard iteration*. However, Newton's iteration method, aimed at finding a solution to the equation $f(x) = 0$, has been known for a longer time.

If F is a contraction, then repeated application of (4.4) gives, for any $n \in \mathbb{N}$,

$$d(y_{n+1}, y_n) \le \alpha^n d(y_1, y_0) \,.$$

Because $0 \le \alpha < 1$, the successive approximations form a sequence of points $y_0, y_1, y_2, \ldots$ in the metric space that cluster together at a rate controlled by α.

The reader may prove that if F is a contraction mapping on M, then F is continuous on M (in the ε-δ definition of continuity, choose $\delta = \varepsilon$). The following is one of the most important theorems in all of mathematics:

Theorem 4.3 (Banach Contraction Mapping Theorem). *Let M be a complete metric space and let $F: M \to M$ be a contraction. Then F has a unique fixed point.*

Proof. Choose any initial point $y_0 \in M$. As above, let $y_{m+1} = F(y_m)$ for all m. For $m > n$,

$$d(y_m, y_n) \leq d(y_m, y_{m-1}) + d(y_{m-1}, y_{m-2}) + \cdots + d(y_{n+2}, y_{n+1}) + d(y_{n+1}, y_n)$$

so that

$$d(y_m, y_n) \leq (\alpha^{m-1} + \alpha^{m-2} + \cdots + \alpha^{n+1} + \alpha^n) d(y_1, y_0)$$

$$= \alpha^n (1 + \alpha + \cdots + \alpha^{m-n-2} + \alpha^{m-n-1}) d(y_1, y_0)$$

$$\leq \left(\frac{\alpha^n}{1 - \alpha} \right) d(y_1, y_0) .$$

Now $\alpha^n/(1 - \alpha)$ can be made arbitrarily small by choosing n large enough. Hence $\{y_m\}$ is a Cauchy sequence. By completeness of M, there is a point $Y \in M$ such that $y_m \to Y$ as $m \to \infty$. By continuity of F,

$$Y = \lim_{m \to \infty} F(y_m) = F\left(\lim_{m \to \infty} y_m \right) = F(Y)$$

and the existence of the fixed point is established. For uniqueness, suppose that $Y = F(Y)$ and $Z = F(Z)$. Then

$$d(Y, Z) = d(F(Y), F(Z)) \leq \alpha d(Y, Z) .$$

But $\alpha < 1$; hence $d(Y, Z) = 0$, and uniqueness is established. $\qquad\square$

We will encounter applications of the contraction mapping theorem in Chap. 5. At this point we present a simple numerical application: the classical iterative formula for the square root of a positive real number.

Example 4.7. Suppose we want to numerically approximate $\sqrt{a}$, where a is a positive real number.

Take any real value $x > 0$ as a first guess for the approximation. If $x < \sqrt{a}$, then $1 < \sqrt{a}/x$ so that $\sqrt{a} < a/x$. Therefore $x < \sqrt{a} < a/x$. On the other hand, suppose $\sqrt{a} < x$. Then $\sqrt{a}/x < 1$ and we have $a/x < \sqrt{a}$. In either case, $\sqrt{a}$ lies between x and a/x and it makes sense to use the average of these numbers as an improved guess at $\sqrt{a}$. We are led to consider the iterative scheme

$$x_{n+1} = \tfrac{1}{2}(x_n + a/x_n) \qquad (n = 1, 2, \ldots) \tag{4.5}$$

which is a very old way of computing $\sqrt{a}$.

In order to apply Theorem 4.3, we must formulate a complete metric space M such that the function

$$f(x) = \tfrac{1}{2}(x + a/x)$$

acts from M into M and is a contraction on M. For M, we will use a real interval of the form $[\xi, \infty)$ with $d(x, y) = |x - y|$.

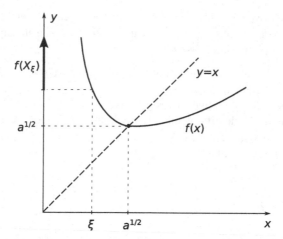

Fig. 4.2 Example on the contraction mapping theorem

Let us write $X_\xi = [\xi, \infty)$ and look for a permissible value of ξ. The minimum value of $f(x)$ occurs at $x = \sqrt{a}$ and is $f(\sqrt{a}) = \sqrt{a}$ (Fig. 4.2). Therefore, if $0 < \xi \leq \sqrt{a}$, then for any $x \geq \xi$ we have $f(x) \geq \sqrt{a} \geq \xi$. In other words, the set image $f(X_\xi) \subseteq X_\xi$. However, we still need $f(x)$ to be a contraction on X_ξ under our chosen metric, and this can further restrict our choice of ξ.

By the mean value theorem we have

$$|f(x) - f(y)| = |f'(\phi)| \, |x - y|$$

with some ϕ lying between x and y. So

$$|f(x) - f(y)| \leq \sup_{\phi \geq \xi} |f'(\phi)| \, |x - y| \, .$$

We see that f is a contraction on X_ξ if $|f'(x)| = |1 - a/x^2|/2 \leq 1/2$ on X_ξ. This inequality is equivalent to

$$-1 \leq 1 - a/x^2 \leq 1 \, .$$

Here the rightmost inequality is valid for any positive x. The leftmost inequality is valid if $x \geq \sqrt{a/2}$, which restricts $\xi \geq \sqrt{a/2}$. Thus, taking $\sqrt{a/2} \leq \xi \leq \sqrt{a}$, we can apply the contraction principle. By the iterative scheme (4.5) with starting point $x_1 \geq \sqrt{a}$, which lies inside X_ξ, we obtain the unique fixed point of f.   □

4.4 Linear Spaces

We know that $\mathbb{R}^n$ and $\mathbb{C}^n$ are vector spaces in which the sum of vectors is defined along with the product of a vector by a scalar, real or complex, depending on the space. Now we extend these ideas to a general case.

A linear space over a field F is a set X, whose elements are called vectors, together with two operations, called vector addition and scalar multiplication, such that the following axioms are satisfied:

1. X is closed with respect to the two operations. That is, $x + y \in X$ and $\alpha x \in X$ whenever $x, y \in X$ and $\alpha \in F$.
2. Addition is both commutative and associative.
3. There is an additive identity element (zero vector) in X. For each vector $x \in X$, there is an additive inverse $-x \in X$.
4. If $x, y \in X$ and $\alpha, \beta \in F$, then:

   (a) $\alpha(x + y) = \alpha x + \alpha y$.
   (b) $(\alpha + \beta)x = \alpha x + \beta x$.
   (c) $(\alpha\beta)x = \alpha(\beta x)$.
   (d) $1x = x$.

If the $F = \mathbb{R}$, then X is a real linear space; if $F = \mathbb{C}$, then X is a complex linear space. We may refer to vectors as elements or even points if this aids in geometrical interpretation.

Example 4.8. $\mathbb{R}^n$ and $\mathbb{C}^n$ are linear spaces. The collection of all real-valued continuous functions on $[a, b]$ is also a linear space.   □

Some terms and concepts are important in the study of linear spaces. A nonempty subset M of a linear space X is a *subspace* of X if M is itself a linear space under the same operations of addition and multiplication as X. A *linear combination* of the vectors $x_1, \ldots, x_n$ is a sum of the form $\alpha_1 x_1 + \cdots + \alpha_n x_n$ where the scalars $\alpha_1, \ldots, \alpha_n \in F$. A set of vectors $\{x_1, \ldots, x_n\}$ is *linearly dependent* if there exist $\alpha_1, \ldots, \alpha_n \in F$, not all zero, such that $\alpha_1 x_1 + \cdots + \alpha_n x_n = 0$. A set of vectors that is not linearly dependent is *linearly independent*. If every vector $x \in X$ can be expressed as a linear combination of the vectors from a set S, then S is a *spanning set* for X. A linearly independent spanning set is a *basis*. A basis is essentially a coordinate system. Any vector space which has a finite spanning set (i.e., any *finite-dimensional space*) contains a basis. All bases of such a space contain the same number of vectors, called the *dimension* of the space. If X has dimension n, then any set of n linearly independent vectors in X is a basis of X.

A *norm* is a real-valued function which assigns to each vector $x \in X$ a number $\|x\|$ such that

N_1. $\|x\| \geq 0$, and $\|x\| = 0$ if and only if $x = 0$;

N_2. $\|\alpha x\| = |\alpha| \, \|x\|$;

N_3. $\|x + y\| \leq \|x\| + \|y\|$.

(On p. 21 we stated the analogous definition for vectors $\mathbf{x} \in \mathbb{R}^n$.) A linear space X, when supplied with a norm $\|\cdot\|$ defined for each element $x \in X$ (which means that its value is a unique finite real number) is a *normed space*. The full notation for the normed space is $(X, \|\cdot\|)$. If the norm is standard or otherwise understood, the notation may be shortened to X.

Theorem 4.4. *A normed space* $(X, \|\cdot\|)$ *is a metric space with the induced or natural metric* $d(x, y) = \|x - y\|$.

Proof. We need only verify satisfaction of the metric axioms. Axiom M_1 is satisfied by virtue of axiom N_1. Axiom M_2 is satisfied because $\|x - y\| = \|y - x\|$. Finally, M_3 is a consequence of N_3 and N_2:

$$d(x, y) = \|x - y\| = \|x - z - (y - z)\| \leq \|x - z\| + \|-(y - z)\| = d(x, z) + d(y, z) .$$

Note that in axiom N_1 we can omit the requirement that $\|x\| \geq 0$, just as we proved that $d(x, y) \geq 0$. □

Because a normed space is a metric space, in cases where an arbitrary Cauchy sequence in a normed space has a limit in the space, we use the term *complete normed space*. Such a space is also termed a *Banach space*, after Stefan Banach, a mathematician who was educated as an engineer, understood the usefulness of such spaces in applications, and developed important aspects of the relevant theory.

Example 4.9. In $\mathbb{R}^n$,

$$\|\mathbf{x}\| = \left(\sum_{k=1}^{n} x_k^2 \right)^{1/2} .$$

A function space is an example of an infinite-dimensional linear space. Norms often used with function spaces include the max norm

$$\|f\| = \max_{x \in [a,b]} |f(x)|$$

(we denoted the space of continuous functions under this norm by $C[a, b]$ and proved that it is a Banach space) and the L_2 norm

$$\|f\| = \left(\int_a^b |f(x)|^2 \, dx \right)^{1/2} .$$

For a particular selection of norm, the function space is defined as the set of all f such that $\|f\| < \infty$. We can also introduce the more general L_p norm:

$$\|f\|_{L_p} = \left(\int_a^b |f(x)|^p \, dx \right)^{1/p} \qquad (p \geq 1) .$$

It is easy to see that this satisfies axioms N_1 and N_2. Satisfaction of N_3 holds by Minkowski's inequality. The space of integrable functions with the L_p norm becomes a Banach space if we use Lebesgue integrals. □

In a normed space X, we have the following result.

Theorem 4.5 (Triangle Inequality). *Assume x, y are two vectors in a normed space. Then*

$$\big|\, \|x\| - \|y\| \,\big| \leq \|x - y\| \leq \|x\| + \|y\| \;.\tag{4.6}$$

Proof. By axiom N_3 with x replaced by $x - y$,

$$\|x\| - \|y\| \leq \|x - y\| \;.$$

Swapping x and y we have, by axiom N_2,

$$\|y\| - \|x\| \leq \|y - x\| = \|(-1)(x - y)\| = \|x - y\| \;.$$

Therefore $\big|\, \|x\| - \|y\| \,\big| \leq \|x - y\|$, and the use of N_3 again yields (4.6). $\square$

Because a normed space is also a metric space, the concepts of Cauchy sequence, convergence, and completeness apply. The following theorem, for instance, is useful in applications.

Theorem 4.6. *In a normed space every Cauchy sequence is bounded.*

Proof. If $\{x_n\}$ is a Cauchy sequence, then with $\varepsilon = 1$ there exists N such that $\|x_n - x_m\| < 1$ whenever $n, m > N$. With $m = N + 1$, this reads $\|x_n - x_{N+1}\| < 1$ whenever $n > N$. For all $n > N$,

$$\|x_n\| = \|x_n - x_{N+1} + x_{N+1}\| \leq \|x_n - x_{N+1}\| + \|x_{N+1}\| < \|x_{N+1}\| + 1 \;.$$

Hence an upper bound for $\|x_n\|$ is $B = \max\{\|x_1\|, \dots, \|x_N\|, \|x_{N+1}\| + 1\}$. $\square$

The reader can extend this proof to show that every Cauchy sequence in a metric space is bounded.

An *inner product* on a real linear space is a function assigning to each pair of vectors x, y a real number $\langle x, y \rangle$ such that:

I_1. $\langle x, y \rangle = \langle y, x \rangle$.

I_2. $\langle \alpha x, y \rangle = \alpha \langle x, y \rangle$.

I_3. $\langle x + y, z \rangle = \langle x, z \rangle + \langle y, z \rangle$.

I_4. $\langle x, x \rangle \geq 0$, and $\langle x, x \rangle = 0$ if and only if $x = 0$.

To define an inner product on a complex linear space, we modify the above so that $\langle x, y \rangle \in \mathbb{C}$, and rewrite I_1 as:

I_1. $\langle x, y \rangle = \overline{\langle y, x \rangle}$.

A linear space furnished with an inner product is an *inner product space*.

Example 4.10. In $\mathbb{R}^n$ and $\mathbb{C}^n$ we use, respectively, the inner product expressions

$$\langle \mathbf{x}, \mathbf{y} \rangle = \sum_{i=1}^{n} x_i y_i \;, \qquad \langle \mathbf{x}, \mathbf{y} \rangle = \sum_{i=1}^{n} x_i \overline{y_i} \;.$$

An inner product of the form

$$\langle f, g \rangle = \int_a^b f(x)\,\overline{g}(x)\,dx$$

is often used with complex-valued functions. □

With inner product structure in a linear space, we can watch more familiar inequalities arise.

Theorem 4.7 (Cauchy–Schwarz Inequality). *Let x, y be two vectors in a complex inner product space. Then*

$$|\langle x, y \rangle| \le \sqrt{\langle x, x \rangle \langle y, y \rangle} \tag{4.7}$$

and equality holds if and only if there is a scalar β such that $x = \beta y$. Furthermore, in the case of equality with $y \ne 0$, $\langle x, y \rangle = \langle \beta y, y \rangle = \beta \langle y, y \rangle$ so $\beta = \langle x, y \rangle / \langle y, y \rangle$. Thus equality holds if and only if $x = 0$ or $y = 0$ or else $x = (\langle x, y \rangle / \langle y, y \rangle) y$.

Proof. By property I_4, for every scalar α we have $0 \le \langle x + \alpha y, x + \alpha y \rangle$ with equality if and only if $x = -\alpha y = \beta y$. By the other properties this inequality can be manipulated into the equivalent form

$$0 \le \langle x, x \rangle + \overline{\alpha}\langle x, y \rangle + \alpha \overline{\langle x, y \rangle} + \alpha \overline{\alpha} \langle y, y \rangle .$$

To shorten the notation, we write $a = \langle x, x \rangle$, $b = \langle x, y \rangle$, $c = \langle y, y \rangle$ and have

$$0 \le |\alpha|^2 c + 2\,\mathrm{Re}[\alpha \overline{b}] + a .$$

Note that a and c are real and nonnegative. If $c \ne 0$, we may put $\alpha = -b/c$ and get $|b|^2 \le ac$ as desired. If $c = 0$ but $a \ne 0$, the roles of x and y may be reversed in the definitions of a, b, c above to yield the same result. If c and a are both zero, then x and y are both zero by I_4, and Cauchy–Schwarz holds trivially. □

Example 4.11. Substituting the expressions given in Example 4.10, we may generate the specific forms

$$\left| \sum_{i=1}^n x_i y_i \right| \le \sqrt{\sum_{i=1}^n x_i^2 \sum_{i=1}^n y_i^2} \,, \qquad \left| \sum_{i=1}^n x_i \overline{y_i} \right| \le \sqrt{\sum_{i=1}^n |x_i|^2 \sum_{i=1}^n |y_i|^2} \,,$$

and

$$\left| \int_a^b f(x)\,\overline{g}(x)\,dx \right| \le \sqrt{\int_a^b |f(x)|^2\,dx \int_a^b |g(x)|^2\,dx} \,.$$

Note the economy of the abstract space approach. □

For any two vectors x, y in a real linear space, the Cauchy–Schwarz inequality can be written as

$$\langle x, y \rangle^2 \leq \langle x, x \rangle \langle y, y \rangle . \tag{4.8}$$

Theorem 4.8 (Minkowski Inequality). *Suppose x, y are two vectors in a linear space. Then*

$$\sqrt{\langle x+y, x+y \rangle} \leq \sqrt{\langle x, x \rangle} + \sqrt{\langle y, y \rangle} . \tag{4.9}$$

Proof.

$$\langle x+y, x+y \rangle = \langle x, x \rangle + 2\,\mathrm{Re}\langle x, y \rangle + \langle y, y \rangle \leq \langle x, x \rangle + 2|\langle x, y \rangle| + \langle y, y \rangle$$

$$\leq \langle x, x \rangle + 2\sqrt{\langle x, x \rangle \langle y, y \rangle} + \langle y, y \rangle = \left(\sqrt{\langle x, x \rangle} + \sqrt{\langle y, y \rangle} \right)^2 .$$

Conditions for equality are treated in Problem 4.8. □

A norm can be induced by the inner product using the equation

$$\|x\|^2 = \langle x, x \rangle . \tag{4.10}$$

The Cauchy–Schwarz and Minkowski inequalities can then be written as

$$|\langle x, y \rangle| \leq \|x\| \|y\| , \qquad \|x + y\| \leq \|x\| + \|y\| ,$$

respectively. Thus we have established

Theorem 4.9. *An inner product space is a normed space with the induced (or natural) norm* (4.10).

Because it is a normed space and therefore a metric space, an inner product space can be complete or incomplete depending on whether all Cauchy sequences have limits in the space. A complete inner product space is termed a *Hilbert space* in honor of David Hilbert.

Example 4.12. On the set of functions continuous on $[a, b]$ with finite a, b, we can introduce the L_2 inner product and its corresponding norm

$$\|f\|_{L_2}^2 = \int_a^b |f(x)|^2 \, dx .$$

Note that the space $C[a, b]$, constructed on the same base set of functions, is a Banach space; with the L_2 norm, however, the set of continuous functions is not complete. This implicitly says that these norms are not equivalent, which can be proved directly. □

In an inner product space we also have the following theorem.

Theorem 4.10 (Parallelogram Law). *Let x, y be vectors in a linear space in which the norm is induced by the inner product. Then*

$$\|x + y\|^2 + \|x - y\|^2 = 2\,\|x\|^2 + 2\,\|y\|^2\ .$$

Proof. This follows from straightforward expansion and manipulation of the quantity $\langle x + y, x + y \rangle + \langle x - y, x - y \rangle$ using the basic inner product properties.   □

Note that in $\mathbb{R}^2$ the vectors x and y represent adjacent sides of a parallelogram, while $x + y$ and $x - y$ are the diagonals.

The following two items are also of some use in applications.

Theorem 4.11 (Continuity of the Inner Product). *Assume the norm is induced by the inner product, and suppose that $x_n \to x$ and $y_n \to y$. Then $\langle x_n, y_n \rangle \to \langle x, y \rangle$.*

Proof. We use the triangle and Cauchy–Schwarz inequalities:

$$
\begin{aligned}
|\langle x_n, y_n \rangle - \langle x, y \rangle| &= |\langle x_n, y_n \rangle - \langle x_n, y \rangle + \langle x_n, y \rangle - \langle x, y \rangle| \\
&= |\langle x_n, y_n - y \rangle + \langle x_n - x, y \rangle| \\
&\leq |\langle x_n, y_n - y \rangle| + |\langle x_n - x, y \rangle| \\
&\leq \|x_n\|\,\|y_n - y\| + \|x_n - x\|\,\|y\|\ .
\end{aligned}
$$

Since $\{x_n\}$ is convergent it is bounded with $\|x_n\| \leq B$ for some finite B. The other n-dependent quantities can be made as small as desired by choosing n sufficiently large.   □

Corollary 4.1 (Continuity of the Norm). *If the norm is induced by the inner product, then $x_n \to x$ implies $\|x_n\| \to \|x\|$.*

Two vectors x, y in an inner product space are *orthogonal* if $\langle x, y \rangle = 0$. We call $\{x_1, x_2, \ldots\}$ an orthogonal set if $\langle x_i, x_j \rangle = 0$ for all $i, j > 0$ with $i \neq j$. An *orthonormal set* is an orthogonal set where $\|x_i\| = 1$ for all i. Mutual orthogonality among the members of a finite set of nonzero vectors $(x_1, \ldots, x_n)$ implies linear independence among those vectors. Indeed, writing out a linear combination, equating it to zero as $\sum_{k=1}^{n} c_k x_k = 0$, and multiplying it by x_i, we find that the only nonzero term on the left is $c_i \langle x_i, x_i \rangle = 0$ and hence $c_i = 0$. As this can be done for any i from 1 to n, we get linear independence of the system. Through the *Gram–Schmidt procedure*, which is produced exactly as in linear algebra for vectors in $\mathbb{R}^n$, we can generate an equivalent mutually orthogonal set $(x_1, x_2, \ldots)$ from any linearly independent set of vectors $(e_1, e_2, \ldots)$ successively:

$$
x_1 = \frac{e_1}{\|e_1\|}\ , \quad x_2 = \frac{e_2 - \langle e_2, x_1 \rangle}{\|e_2 - \langle e_2, x_1 \rangle\|}\ , \quad \ldots\ , \quad x_k = \frac{e_k - \sum_{j=1}^{k-1}\langle e_k, x_j \rangle}{\left\|e_k - \sum_{j=1}^{k-1}\langle e_k, x_j \rangle\right\|}\ , \quad \ldots\ .
$$

It is interesting to note that this process, while theoretically perfect, is troublesome in practice when applied to large sets of vectors because of the way errors accumulate in numerical calculations.

A handy theorem, which follows from the inner product and orthogonality definitions, is the following:

Theorem 4.12 (Pythagorean Theorem). *Let the vectors x and y be orthogonal. Then*

$$\|x + y\|^2 = \|x\|^2 + \|y\|^2 \ .$$

The proof is left to the reader.

Orthogonal Projection and Expansion

Working with geometry in $\mathbb{R}^3$, we often use the notion of orthogonality because it can yield relatively simple optimization methods such as projections onto axes. We extend this to an abstract Hilbert space H, where there is also a notion of orthogonality based on the inner product.

Suppose a vector $x \in H$ lies outside M, a closed subspace of H. The case when $x \in M$ is trivial for the problem considered below. Some optimization schemes require a vector $m \in M$ that is "closest" to x in the sense that $\|x - m\|$ is minimized. Such a vector m_0 is known as a *minimizing vector*.

Theorem 4.13. *Let x be an element of a Hilbert space H and M a closed subspace of H. In M, there exists the best approximation m_0 of x:*

$$\|x - m_0\| = \inf_{m \in M} \|x - m\| \ .$$

The minimizing element m_0 is unique. The difference $x - m_0$ is orthogonal to M, i.e., it is orthogonal to any $m \in M$ so that $\langle x - m_0, m \rangle = 0$.

Proof. First we wish to show that, corresponding to the given x, there exists $m_0 \in M$ such that for all $m \in M$,

$$\|x - m\| \geq \|x - m_0\| \ .$$

Let $x \in H$ be given. If $x \in M$, we simply choose $m_0 = x$. If $x \notin M$, we define

$$\delta = \inf_{m \in M} \|x - m\| \ .$$

Note that for any $m_i, m_j \in M$,

$$\|m_j - m_i\|^2 = \|(m_j - x) + (x - m_i)\|^2$$

and by Theorem 4.10

$$\|(m_j - x) + (x - m_i)\|^2 + \|(m_j - x) - (x - m_i)\|^2 = 2\|x - m_j\|^2 + 2\|x - m_i\|^2 \ .$$

Hence

$$\|m_j - m_i\|^2 = 2\|x - m_j\|^2 + 2\|x - m_i\|^2 - 4\left\|x - \frac{m_i + m_j}{2}\right\|^2$$

$$\le 2\|x - m_j\|^2 + 2\|x - m_i\|^2 - 4\delta^2. \tag{4.11}$$

The inequality follows by definition of δ, because M is a subspace and therefore contains the vectors $(m_i + m_j)/2$. Now let $\{m_i\}$ be a sequence in M such that $\|x - m_i\| \to \delta$. As $i, j \to \infty$ then, the squeeze principle gives $\|m_j - m_i\| \to 0$ so that $\{m_i\}$ is a Cauchy sequence in M (and in H). Because $\{m_i\}$ is a Cauchy sequence and M is closed, $\{m_i\}$ converges to a point $m_0 \in M$. By continuity of the norm, $\|x - m_0\| = \delta$. The minimizing vector is unique. For supposing that $m_{01}, m_{02} \in M$ are two minimizing vectors, then choosing $m_i = m_{01}$ and $m_j = m_{02}$ in (4.11) gives

$$\|m_{02} - m_{01}\|^2 \le 2\|x - m_{02}\|^2 + 2\|x - m_{01}\|^2 - 4\delta^2 \le 2\delta^2 + 2\delta^2 - 4\delta^2 = 0,$$

and hence $m_{02} = m_{01}$.

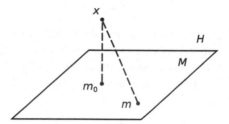

Fig. 4.3 Minimizing vector

From an intuitive "best approximation" standpoint, it is not surprising (Fig. 4.3) that m_0 is the unique minimizing vector if and only if the error vector $x - m_0$ is orthogonal to every $m \in M$. As the inner product is linear with respect to the first argument and conjugate-linear with respect to the second argument, it is sufficient to prove that $x - m_0$ is orthogonal to all unit vectors m of M. Supposing the existence of a normalized $m \in M$ that is not orthogonal to $x - m_0$, we would, contrary to the infimum property of m_0, get an element in M closer to x. Indeed, denote $\langle x - m_0, m \rangle = \alpha \ne 0$. The vector $m_0 + \alpha m$ satisfies

$$\|x - (m_0 + \alpha m)\|^2 = \langle (x - m_0) - \alpha m, (x - m_0) - \alpha m \rangle$$

$$= \langle x - m_0, x - m_0 \rangle - \langle \alpha m, x - m_0 \rangle - \langle x - m_0, \alpha m \rangle + \langle \alpha m, \alpha m \rangle$$

$$= \|x - m_0\|^2 - 2\,\mathrm{Re}(\langle \alpha m, x - m_0 \rangle) + |\alpha|^2 \|m\|^2$$

$$= \|x - m_0\|^2 - 2|\alpha|^2 + |\alpha|^2$$

$$= \|x - m_0\|^2 - |\alpha|^2$$

$$< \|x - m_0\|^2 .$$

Because $\|x - (m_0 + \alpha m)\| < \|x - m_0\|$, we see that m_0 cannot be a minimizing vector—a contradiction.

The orthogonality of $x - m_0$ to every $m \in M$ suffices for uniqueness of m_0. Indeed, for any $m \in M$ we have

$$\|x - m\|^2 = \|x - m_0 + m_0 - m\|^2 = \|x - m_0\|^2 + \|m - m_0\|^2 \ ,$$

hence $\|x - m\| > \|x - m_0\|$ unless $m = m_0$. Note that the proof is also valid in a real Hilbert space H.   □

Abstract Fourier Series

Let f be an arbitrary vector in a Hilbert space H, and let $\{x_1, x_2, \ldots\}$ be an orthonormal set in the space. For fixed $n \in \mathbb{N}$, form the subspace M_n generated by all linear combinations

$$g = \sum_{i=1}^{n} c_i x_i$$

with any scalars c_i. We can apply Theorem 4.13 on the best approximation of f by an element from M_n. The theorem provides us with a unique element g_0 but now, using the properties of the inner product and orthonormality of set $\{x_1, \ldots, x_n\}$ we can construct the minimizer explicitly. Denoting by $\alpha_k = \langle f, x_k \rangle$ the kth Fourier coefficient of f, we have

$$\left\| f - \sum_{k=1}^{n} c_k x_k \right\|^2 = \left\langle f - \sum_{k=1}^{n} c_k x_k, f - \sum_{k=1}^{n} c_k x_k \right\rangle$$

$$= \langle f, f \rangle - \left\langle f, \sum_{k=1}^{n} c_k x_k \right\rangle - \left\langle \sum_{k=1}^{n} c_k x_k, f \right\rangle + \left\langle \sum_{k=1}^{n} c_k x_k, \sum_{m=1}^{n} c_m x_m \right\rangle$$

$$= \langle f, f \rangle - \sum_{k=1}^{n} \overline{c_k} \alpha_k - \sum_{k=1}^{n} c_k \overline{\alpha_k} + \sum_{k=1}^{n} c_k \overline{c_k}$$

$$= \|f\|^2 - \sum_{k=1}^{n} |\alpha_k|^2 + \sum_{k=1}^{n} |c_k - \alpha_k|^2 \ . \tag{4.12}$$

The term $\|f\|^2 - \sum_{k=1}^{n} |\alpha_k|^2$ is fixed; hence, with $c_k = \alpha_k$ the function $f - g$ of the variables $c_1, \ldots, c_n$ is minimized in the least squares sense in H. We get the needed minimizer explicitly:

$$g_0 = \sum_{k=1}^{n} \alpha_k x_k \ ,$$

which is a portion of the Fourier series for f.

From (4.12), we deduce that for Fourier series *Bessel's inequality*

$$\sum_{i=k}^{n} |\langle f, x_k \rangle|^2 \leq \|f\|^2$$

holds. As $n \to \infty$, the resulting series on the left must converge (by the theorem that a numerical series $\sum_{k=1}^{\infty} a_k$ with $a_k \geq 0$, whose partial sums $S_n = \sum_{k=1}^{n} a_k \leq B$ with a constant B independent of n, must converge); therefore, we deduce the *Riemann lemma*

$$\lim_{k \to \infty} \langle f, x_k \rangle = 0 .$$

We have established some facts on Fourier series in a complex Hilbert space H. It is even easier to prove them in a real space H.

Most interesting in applications is the question when we actually have

$$f = \sum_{k=1}^{\infty} \langle f, x_k \rangle x_k . \tag{4.13}$$

A sufficient condition is the following. Let $\{x_1, x_2, \ldots\}$ be an orthonormal set such that for any $f \in H$ the set of equalities $\langle f, x_k \rangle = 0$ for all $k = 1, 2, \ldots$ implies $f = 0$. Then (4.13) holds and so does *Parseval's equality*

$$\sum_{k=1}^{\infty} |\langle f, x_k \rangle|^2 = \|f\|^2 .$$

Example 4.13. In the classical setting for Fourier series, H is the set of square-summable functions on $[0, 2\pi]$ using Lebesgue integration with the inner product

$$\langle f, g \rangle = \frac{1}{2\pi} \int_0^{2\pi} f(x)\overline{g}(x) \, dx .$$

Let M be the subspace generated by all trigonometric polynomials of a fixed order n. That is, let

$$\{x_1, \ldots, x_n\} = \{1, e^{\pm ix}, e^{\pm 2ix}, \ldots, e^{\pm mix}\} .$$

The Fourier coefficients c_i chosen as above makes $g(x)$ the minimizing vector, the vector closest to a given f. $\square$

In applications, orthonormal series are typically encountered in two ways: (1) as orthogonal polynomials arising from certain definitions, and (2) as eigenfunctions of boundary value problems. The latter, in turn, arise as intermediate problems when certain self-adjoint boundary value problems for linear partial differential equations are solved by separation of variables.

4.5 Operators

We know that a function f is a relation between two sets. To each point of the domain of f there corresponds just one point of the range of f. For an ordinary function, the domain and range lie in $\mathbb{R}$. For a multivariable function, the domain lies in $\mathbb{R}^n$ and the range can be in $\mathbb{R}^m$; if $m > 1$, then f is called a vector function. In this way we can consider the relation

$$\mathbf{y} = A\mathbf{x} \, ,$$

where A is an $n \times n$ matrix and $\mathbf{x}$ and $\mathbf{y}$ are n-dimensional column vectors, as a function from $\mathbb{R}^n$ to $\mathbb{R}^n$.

It is often advantageous to consider a function as a whole object in an abstract space (e.g., as a "point" in a metric space or an "element" in a normed space). We can consider how this object is mapped into objects in other spaces. The derivative relationship

$$\frac{d}{dx} \sin x = \cos x \, ,$$

for example, can be regarded as mapping a function continuously differentiable on $[a, b]$ (i.e., $\sin x$) into a function continuous on $[a, b]$ (i.e., $\cos x$). We could consider pointwise behavior as usual (i.e., the values the functions take for various $x \in [a, b]$), but instead come to picture these functions as elements of function spaces. The points of one space are mapped to points of the other space by the differentiation operator. An *operator* A is a relation between the points of its domain $D(A)$ and its range $R(A)$ such that to each point of $D(A)$ there corresponds only one point of $R(A)$. Here we have essentially repeated the definition of a function without specifying the sets $D(A)$ and $R(A)$ in a concrete fashion. If these sets lie in $\mathbb{R}^n$ and $\mathbb{R}^m$, respectively, we get an ordinary vector functions in n variables. But the use of other spaces calls for the term operator. A compact notation used for an operator A is

$$A \colon D(A) \subset X \to R(A) \subset Y \, ,$$

where X, Y can be metric spaces, normed spaces, inner product spaces, or other types of spaces (such as topological spaces). So we consider an operator as having three parts: a domain $D(A)$, a range $R(A)$, and a mapping rule between these two sets. An alteration in any of these parts yields a different operator. However, if we restrict or extend $D(A)$, and change $R(A)$ accordingly, we call such an operator a *restriction* or an *extension* of A to the respective domain. An operator A having $R(A)$ in $\mathbb{R}$ or $\mathbb{C}$ is termed a *functional*. Hence we can refer to metrics, norms, and inner products as functionals acting in the corresponding spaces.

Example 4.14. A functional is partly specified by the mapping rule

$$F(f) = \int_a^b f(x) \, dx \, .$$

Here we can assign $D(F) = C[a, b]$. We get an extension of F by assigning $D(A)$ as the set of functions that are merely piecewise continuous on $[a, b]$. Another extension would be obtained by making $D(A)$ the set of all functions that are integrable on $[a, b]$.   □

Example 4.15. The derivative operation given by

$$A(f) = \frac{d}{dx} f(x)$$

defines an operator if we specify a suitable domain. We can say that $D(A) = C^{(1)}[a, b]$, the space of functions that are continuously differentiable on $[a, b]$; the corresponding range is $R(A) = C[a, b]$. An extension of the resulting operator could be obtained by changing $D(A)$ to the set of all functions that are merely differentiable on $[a, b]$ (not necessarily continuously differentiable).   □

Analogously to what is done in calculus, we can introduce the idea of continuity of an operator A. We will suppose that $D(A) \subset X$ and $R(A) \subset Y$ where X and Y are normed spaces. We say that A is continuous at $x^* \in X$ if for any $\varepsilon > 0$ there exists $\delta > 0$, depending on ε, such that $\|A(x) - A(x^*)\|_Y < \varepsilon$ whenever $\|x - x^*\|_X < \delta$. We call A continuous in a set $S \subseteq D(A)$ if it is continuous at each point of S.

Example 4.16. The matrix operator $A \colon \mathbb{R}^n \to \mathbb{R}^n$ given by $A(\mathbf{x}) = A\mathbf{x}$ is continuous on $\mathbb{R}^n$ if all the elements of the matrix A are finite. That is why continuity does not arise as an interesting topic in ordinary linear algebra.   □

The question of continuity of an operator acting between normed spaces is more complicated than the corresponding question in calculus. We will consider it relative to a special class of operators.

Linear Operators

An operator $A \colon D(A) \subset X \to Y$, where X and Y are normed spaces, is a *linear operator* if for any $x_1, x_2 \in X$ and scalars c_1, c_2 we have

$$A(c_1 x_1 + c_2 x_2) = c_1 A(x_1) + c_2 A(x_2) .$$

In the notation for linear operators, we often omit the parentheses and write $A(x) \equiv Ax$ as is customary for matrix operators. Examples 4.14–4.16 featured linear operators. Note that the elementary linear function $y = kx + b$ can be considered as a linear operator if and only if $b = 0$.

It is easy to see that a linear operator continuous at zero is continuous at any point. For continuity of a linear operator A at zero, it is sufficient that there is a constant k such that for all $x \in X$ we have

$$\|Ax\|_Y \le k \|x\|_X . \tag{4.14}$$

As it is clear which spaces pertain to x and Ax, we will omit the subscripts Y, X and write this as $\|Ax\| \leq k\|x\|$. If such a constant k exists, then A is called a *bounded operator*.

Theorem 4.14. *A linear operator* $A\colon X \to Y$ *is continuous on* X *if and only if it is bounded.*

Proof. We must show that if A is continuous at zero then it is bounded. So let A be continuous at zero. As $A0 = 0$, we have, by definition of continuity, that for $\varepsilon = 1$ there is a $\delta > 0$ such that $\|Ax - 0\| = \|Ax\| < 1$ whenever $\|x\| < \delta$. Take any nonzero $x \in X$; then $\|\delta x/(2\|x\|)\| = \delta/2$. Thus

$$\left\| A\frac{\delta x}{2\|x\|} \right\| < 1$$

and by linearity of A we get $\|Ax\| < (2/\delta)\|x\|$. This means that A is bounded and we can take the constant $k = 2/\delta$.   □

A practical way to verify continuity for a linear operator A is to verify boundedness.

Example 4.17. Let $Af = \int_a^b f(x)\,dx$ define a linear functional in $C[a,b]$. Then

$$|Af| = \left| \int_a^b f(x)\,dx \right| \leq \max_{x \in [a,b]} |f(x)| \int_a^b dx = (b-a)\|f\|_{C[a,b]} \ .$$

Equality holds when $f(x) = 1$, so the constant $k = b - a$ cannot be improved.   □

Example 4.18. Fredholm's integral operator

$$(Af)(x) = \int_a^b K(x,s)f(s)\,ds \ , \tag{4.15}$$

where $K(x,s)$ is a function in two variables x and s, is a linear operator from $C[a,b]$. It acts to $C[a,b]$ if K is a continuous function on $[a,b] \times [a,b]$. The estimate

$$\max_{x \in [a,b]} \left| \int_a^b K(x,s)f(s)\,ds \right| \leq \max_{s \in [a,b]} |f(s)| \cdot \max_{x \in [a,b]} \int_a^b |K(x,s)|\,ds$$

shows that it is bounded and hence continuous in $C[a,b]$ with constant

$$k = \max_{x \in [a,b]} \int_a^b |K(x,s)|\,ds \ ,$$

as $K(x,s)$ is a continuous function.   □

Norm of a Linear Operator. Similar to the norm of a matrix, we can introduce the norm of a continuous operator $A\colon X \to Y$: it is smallest of the constants k in (4.14). So for the norm of A, denoted $\|A\|$, we have

$$\|Ax\| \leq \|A\|\,\|x\| \ ,$$

and for any $\varepsilon > 0$ there is $x^* \in X$ such that

$$\|Ax^*\| > \|A - \varepsilon\| \|x^*\| .$$

It can be shown that the set of all continuous linear operators from X to Y constitutes a normed space with this operator norm. The space is denoted $L(X, Y)$.

Note that we can use different norms on the spaces X and Y. The norm of an operator $A: X \rightarrow Y$ can change accordingly. Clearly it also changes if we extend or restrict the domain of the operator. Alterations in the domain or range used with a given mapping rule can make the resulting operator continuous or discontinuous. A good example of this is the derivative action d/dx. If we consider it as an operator from $C^{(1)}[a, b]$, the space of continuously differentiable functions, to $C[a, b]$, then it is continuous. Taking the standard norm in $C^{(1)}[a, b]$, which is

$$\|f\| = \max_{x \in [a,b]} |f(x)| + \max_{x \in [a,b]} |f'(x)| ,$$

we can show that its norm is unity. If we consider it acting in the space $C[a, b]$ (i.e., from $C[a, b]$ to $C[a, b]$), we get a discontinuous operator. We could show the nonexistence of a constant k, but here it is simpler to see that there are continuous functions that are not differentiable at certain points. For these, the norm inequality cannot exist.

Finally, consider the equation

$$x = Ax + b$$

with a linear operator A acting in a normed space X (from X to X) and $b \in X$. If $\|A\| < 1$, we can use Banach's iteration scheme to solve this equation:

$$x_{k+1} = Ax_k + b$$

with any initial approximation. Indeed, the operator $Bx = Ax + b$ is a contraction operator:

$$\|Bx - By\| = \|Ax - Ay\| = \|A(x - y)\| \leq \|A\| \|x - y\| .$$

4.6 Problems

4.1. Prove that in any metric space

(a) $|d(x, y) - d(x, z)| \leq d(y, z)$,

(b) $d(x_1, x_n) \leq d(x_1, x_2) + d(x_2, x_3) + \cdots + d(x_{n-1}, x_n)$.

4.2. Show that any metric $d(x, y)$ is a continuous function in both of its arguments.

4.3. Prove that if a sequence converges in a metric space, then its limit is unique.

4.4. (The space ℓ_p.) Let $p \geq 1$ be a fixed real number, let X be the set of all real sequences of the form $\mathbf{x} = \{\xi_1, \xi_2, \ldots\}$ such that $\sum_{i=1}^{\infty} |\xi_i|^p$ is convergent. For two points $\mathbf{x} = \{\xi_1, \xi_2, \ldots\}$ and $\mathbf{y} = \{\eta_1, \eta_2, \ldots\}$, let the distance be defined as

$$d(\mathbf{x}, \mathbf{y}) = \left(\sum_{i=1}^{\infty} |\xi_i - \eta_i|^p \right)^{1/p}.$$

Show that

(a)   the series defining $d(\mathbf{x}, \mathbf{y})$ is convergent for all $\mathbf{x}, \mathbf{y} \in X$, and

(b) X is a metric space.

4.5. Show that the following are metric spaces:

(a)   the set of functions continuous on $[a, b]$ with distance defined using

$$d(f, g) = \int_a^b |f(x) - g(x)| \, dx \, ;$$

(b)   the set of all bounded sequences $\{x_i\}$ with

$$d(x, y) = \sup_{1 \leq i < \infty} |x_i - y_i| \, .$$

4.6. Show that (4.10) generates a norm.

4.7. In a triangle ABC let the points β and α be the midpoints of sides AC and BC, respectively. Show that if the side lengths of the triangle satisfy $AC > BC$, then the medians satisfy $A\alpha > B\beta$.

4.8. State and prove conditions for equality in Minkowski's inequality.

4.9. Use the Cauchy–Schwarz inequality to prove Theorem 1.4.

4.10. Show that the Fredholm operator of (4.15) is bounded in the set of continuous functions with the norm of $L_2(a, b)$.

4.11. Let $\{x_n\}$ be a sequence of points in a metric space (X, d). Show that if $d(x_n, x_{n+1}) < 2^{-n}$ for $n = 1, 2, \ldots$, then $\{x_n\}$ is a Cauchy sequence.

4.12. Show that in a normed space the following limit theorems hold.

(a)   If $x_m \to x$ and $y_m \to y$, then $x_m + y_m \to x + y$ as $m \to \infty$.

(b)   If $x_m \to x$ and $\lambda_m \to \lambda$, then $\lambda_m x_m \to \lambda x$.

Chapter 5
Some Applications

5.1 Introduction

The reader who has worked patiently through the mathematical content of the previous chapters should be comfortable dealing with the applications treated here. These topics were chosen for variety and are presented in no particular order (just as we might encounter them in practice).

5.2 Estimation of Integrals

This idea was introduced in Chap. 2, and we now offer some additional examples. Note that the triangle, Cauchy–Schwarz, and Minkowski inequalities provide upper bounds for an integral. A useful lower bound can often be obtained from the Chebyshev inequality.

Bounds for integrals are required, for example, when we attempt to estimate the norm of a Fredholm integral operator. A rough estimate of a norm will sometimes suffice; but if we wish to apply Banach's iteration procedure, however, we need the best possible estimate. It may be possible to obtain an approximation using direct numerical calculation on a computer with interval control. Here we consider some simple examples of estimation techniques.

Example 5.1. Consider the integral

$$I = \int_0^1 (1 + x^5)^{1/2} \, dx.$$

Because the integrand takes extreme values of 1 and $\sqrt{2}$ on $[0, 1]$, the inequality $1 \le I \le \sqrt{2}$ is easily obtained. However, the Cauchy–Schwarz inequality with $g(x) \equiv 1$ gives an improved upper bound:

$$I \le \left(\int_0^1 (1 + x^5)\, dx \right)^{1/2} = (7/6)^{1/2} \approx 1.0801.$$

So $1 \le I \le (7/6)^{1/2}$. Numerical evaluation with interval software gives $I \approx 1.07467$. We mentioned interval analysis in Example 1.7; an introduction to the subject appears in Chap. 7.   □

Example 5.2. Given two functions $f(t)$ and $g(t)$ for $t \in (-\infty, \infty)$, the convolution of $f(t)$ and $g(t)$, written $f(t) * g(t)$, is defined by

$$f(t) * g(t) = \int_{-\infty}^{\infty} f(x)g(t - x)\, dx$$

provided the integral exists. The function $f * g$ is bounded if $f(t)$ and $g(t)$ are square integrable on $(-\infty, \infty)$ (see Problem 3.18). For by the Cauchy–Schwarz inequality, we have

$$|f(t) * g(t)|^2 \le \int_{-\infty}^{\infty} |f(x)|^2\, dx \int_{-\infty}^{\infty} |g(t - x)|^2\, dx.$$

Hence $|f(t) * g(t)| < \infty$ for all t, and $f * g$ is bounded.   □

Example 5.3. The integrand of

$$I = \int_2^5 e^x(x + 1)\, dx$$

is a product of functions that increase on $[2, 5]$. Hence by the Chebyshev inequality

$$I \ge \frac{1}{3} \int_2^5 e^x dx \int_2^5 (x + 1)\, dx = \tfrac{9}{2}(e^5 - e^2) \approx 635.$$

An upper bound can be obtained from the Cauchy–Schwarz inequality:

$$I \le \left(\int_2^5 e^{2x}\, dx \int_2^5 (x + 1)^2\, dx \right)^{1/2} \approx 832.$$

The precise value of I to eight places via interval computation is 727.28768.   □

5.3 The o and O Symbols

Sometimes it is only necessary to know the order of growth or decrease of a function near a point. The order symbols O and o permit us to compare an object of interest with a class of test objects.

If f and g are functions of x, we say that

$$f(x) = o(g(x)) \quad \text{when } x \to x_0 \tag{5.1}$$

if

$$\lim_{x \to x_0} \frac{f(x)}{g(x)} = 0.$$

For example, we have $x = o(x^2)$ when $x \to \infty$. We sometimes say that f is of smaller order than g as $x \to x_0$, but this does not imply that the functions in question tend toward zero (in our example both x and x^2 tend to ∞ as $x \to \infty$).

We say that

$$f(x) = O(g(x)) \quad \text{when } x \to x_0 \tag{5.2}$$

if there is a constant B such that $|f(x)/g(x)| \le B$ in some neighborhood of x_0. We sometimes say that $f(x)$ and $g(x)$ are of the same order of magnitude at x_0. In this case the existence of a limit is not required; for example, we have $\sin(1/x) = O(1)$ when $x \to 0$ (which is simply to say that $\sin(1/x)$ is bounded in any neighborhood of $x = 0$).

If (5.1) holds, then so does (5.2); indeed, (5.1) means that $|f(x)/g(x)| < \varepsilon$ for x sufficiently close to x_0. In this sense then, if statement (5.1) holds, it provides more information (or is sharper) than statement (5.2).

Normally the order symbols are used when we know the principal part of the behavior of a function near x_0 and wish to say that the remaining part, which we did not get precisely, is small in comparison. The statement

$$f(x) = g(x) + o(h(x)) \quad \text{when } x \to x_0$$

means that $f(x) - g(x) = o(h(x))$ when $x \to x_0$. Here $g(x)$ is said to be the principal part of $f(x)$ when $x \to x_0$. The statement

$$f(x) = g(x) + O(h(x)) \quad \text{when } x \to x_0$$

is interpreted analogously.

Example 5.4. When $|x| < 1$, the Taylor expansion for $\cos x$ is an alternating series with terms having absolute values that decrease monotonically to zero. In this case we may use Leibnitz's theorem to estimate the difference

$$D(x) = \left| \cos x - \sum_{k=0}^{n} \frac{(-1)^n x^{2n}}{(2n)!} \right| < \frac{|x|^{2n+2}}{(2n+2)!}.$$

If we are interested only in the order of decay of $D(x)$ near $x = 0$, then either of the statements

$$D(x) = O(|x|^{2n+2}) \quad \text{or} \quad D(x) = o(|x|^{2n}) \quad \text{when } x \to 0$$

may be used. □

To show how the use of o and O can provide statements of varying precision about a function, let us take $f(x) = e^x$ and $x_0 = 0$. We may compare sharpness (for expansions of the same function) if the terms before the order symbols are the same.

Here the sharper expansion is the one that provides o or O terms of higher order. The following expansions of e^x near $x = 0$ are listed in order of increasing sharpness:

1. $e^x = 1 + x + o(x)$ when $x \to 0$,
2. $e^x = 1 + x + O(x^2)$ when $x \to 0$.
3. $e^x = 1 + x + x^2/2 + o(x^2)$ when $x \to 0$.

The order symbols often appear in asymptotic analysis. We will introduce the idea of an asymptotic expansion in the next section.

5.4 Series Expansions

Series of functions arise in many contexts. Suppose the functions $u_i(x)$ ($i \in \mathbb{N}$) have a common domain D along the x-axis. The nth partial sum of the series $\sum_{i=1}^{\infty} u_i(x)$ is

$$S_n(x) = \sum_{i=1}^{n} u_i(x).$$

We say that $\sum_{i=1}^{\infty} u_i(x)$ *converges uniformly* to $u(x)$ on D if and only if for every $\varepsilon > 0$ there is an $N > 0$ (dependent on ε but not on x) such that $|u(x) - S_n(x)| < \varepsilon$ whenever $n > N$ and $x \in D$. Note that if $D = [a, b]$, then uniform convergence is the convergence of the sequence $\{S_n(x)\}$ to $u(x)$ in the space $C[a, b]$. Uniform convergence can settle the question whether a given series of functions can be integrated or differentiated termwise. The manipulation

$$\int_a^b \sum_{i=1}^{\infty} u_i(x)\, dx = \sum_{i=1}^{\infty} \int_a^b u_i(x)\, dx$$

is valid if the functions $u_i(x)$ are integrable and $\sum_{i=1}^{\infty} u_i(x)$ converges uniformly on $[a, b]$. We have

$$\frac{d}{dx} \sum_{i=1}^{\infty} u_i(x) = \sum_{i=1}^{\infty} \frac{d}{dx} u_i(x)$$

for all $x \in [a, b]$ if the functions $u_i(x)$ have continuous derivatives in $[a, b]$, the series $\sum u_i(x)$ converges uniformly on $[a, b]$, and the differentiated series on the right converges uniformly in $[a, b]$.

A useful lemma called the *Weierstrass M-test* provides sufficient conditions for uniform convergence. Suppose a convergent series of positive constants $\sum_{i=1}^{\infty} M_i$ can be found such that $|u_i(x)| \le M_i$ for all i and all $x \in D$. Call $u(x) = \sum_{i=1}^{\infty} u_i(x)$ and $M = \sum_{i=1}^{\infty} M_i$. Then

$$|u(x) - S_n(x)| = \left| \sum_{i=n+1}^{\infty} u_i(x) \right| \le \sum_{i=n+1}^{\infty} |u_i(x)| \le \sum_{i=n+1}^{\infty} M_i = \left| M - \sum_{i=1}^{n} M_i \right|.$$

But $\sum_{i=1}^{\infty} M_i$ converges; hence, given $\varepsilon > 0$, we can choose N such that for $n > N$ the last quantity is less than ε. So $\sum_{i=1}^{\infty} u_i(x)$ converges uniformly on D.

Example 5.5. Let $f(x)$ be periodic with period 2π. The series

$$a_0 + \sum_{n=1}^{\infty} a_n \cos nx + b_n \sin nx$$

is the Fourier series of $f(x)$ if

$$a_0 = \frac{1}{2\pi} \int_{-\pi}^{\pi} f(x)\,dx,$$

and for $n \in \mathbb{N}$,

$$a_n = \frac{1}{\pi} \int_{-\pi}^{\pi} f(x) \cos nx\,dx, \qquad b_n = \frac{1}{\pi} \int_{-\pi}^{\pi} f(x) \sin nx\,dx.$$

Convergence (especially uniform convergence) of Fourier series has received a great deal of study, and a general treatment of the topic involves Lebesgue integration. However, it is instructive to see a simple set of convergence conditions established using the M-test. If $f(x)$ has continuous derivatives through order two for all x, then the trigonometric Fourier series of $f(x)$ converges uniformly everywhere. We integrate the formula for a_n by parts twice and make use of the periodicity of $f(x)$ and its derivative to get

$$a_n = -\frac{1}{n^2\pi} \int_{-\pi}^{\pi} f''(x) \cos nx\,dx.$$

Now since $f''(x)$ is continuous on $[-\pi, \pi]$, it attains maximum and minimum values on that interval. Hence, for some $B > 0$,

$$|a_n| \le \frac{1}{n^2\pi} \int_{-\pi}^{\pi} |f''(x)|\,dx \le \frac{2B}{n^2}.$$

Similarly, $|b_n| \le 2B/n^2$. The desired conclusion follows from the M-test and convergence of the numerical series $\sum n^{-p}$ for $p = 2 > 1$. Note that we have established convergence of the Fourier series but not the fact that the series sums to $f(x)$. However, convergence in the sense of $L_2(-\pi, \pi)$ is ensured by the general results for abstract Fourier series if $f \in L_2(-\pi, \pi)$. $\square$

Another important type of expansion is the asymptotic expansion. In an *asymptotic sequence* of functions each term is dominated, in o fashion, by the previous term. Thus, $\{w_n(x)\}$ is an asymptotic sequence for $x \to x_0$ if $w_{n+1}(x) = o(w_n(x))$ or, equivalently,

$$\lim_{x \to x_0} \frac{w_{n+1}(x)}{w_n(x)} = 0.$$

The weighted sum $\sum a_n w_n(x)$, where the a_n are constants, might turn out to be a good approximation to some function $f(x)$ when x is close to x_0. If

$$f(x) - \sum_{n=1}^{m} a_n w_n(x) = o(w_m(x)) \qquad (x \to x_0), \qquad (5.3)$$

then the summation is an *asymptotic expansion* to m terms of $f(x)$ for $x \to x_0$, and we write

$$f(x) \sim \sum_{n=1}^{m} a_n w_n(x) \qquad (x \to x_0).$$

The special case $m = 1$ gives rise to a single-term *asymptotic formula* for $f(x)$. For fixed m, the difference between $f(x)$ and its asymptotic expansion approaches zero faster than the last term included in the expansion. A special case of an asymptotic expansion at $x = x_0$ is provided by the Taylor expansion of $f(x)$ in a power series with respect to $(x - x_0)^k$ when f has only a finite number of continuous derivatives at x_0.

Many functions have asymptotic expansions for large x of the form

$$f(x) \sim \sum_{n=0}^{m} \frac{a_n}{x^n} \qquad (x \to \infty),$$

i.e., in inverse powers of x. If such a function can be written without approximation as

$$f(x) = \sum_{n=0}^{m} \frac{a_n}{x^n} + R_m(x),$$

then a suitable criterion on the remainder term $R_m(x)$ is that for any fixed m we have

$$R_m(x) = O\left(\frac{1}{x^{m+1}}\right) \text{ as } x \to \infty. \qquad (5.4)$$

In other words there is a finite B such that for sufficiently large x,

$$|R_m| \leq B/x^{m+1}.$$

Hence $x^m |R_m| \leq B/x$, and the quantity $R_m/(1/x^m)$ is squeezed to zero as $x \to \infty$, implementing the o requirement in (5.3).

Example 5.6. Consider the function $g(x)$ defined by the integral

$$g(x) = \int_x^{\infty} \frac{e^{x-t}}{t} dt.$$

An m-fold integration by parts yields

$$g(x) = \frac{1}{x} - \frac{1}{x^2} + \frac{2!}{x^3} - \frac{3!}{x^4} + \cdots + (-1)^{m-1} \frac{(m-1)!}{x^m} + R_m(x),$$

where

$$R_m(x) = (-1)^m m! \int_x^\infty \frac{e^{x-t}}{t^{m+1}} \, dt.$$

But we can write

$$\int_x^\infty \frac{e^{x-t}}{t^{m+1}} \, dt \le \int_x^\infty \frac{e^{x-t}}{x^{m+1}} \, dt = \frac{1}{x^{m+1}}$$

so that (5.4) is satisfied, and thus

$$g(x) \sim \frac{1}{x} - \frac{1}{x^2} + \frac{2!}{x^3} - \frac{3!}{x^4} + \cdots + (-1)^{m-1} \frac{(m-1)!}{x^m} \quad \text{as } x \to \infty.$$

So we have established the mth order asymptotic expansion for the dependence of the integral on the parameter x. We should note that this situation is typical. The series

$$\sum_{m=1}^\infty (-1)^{m-1} \frac{(m-1)!}{x^m}$$

diverges at any value of x, no matter how large. However, we can use the asymptotic expansion to approximate the value of the integral, and we know the precision of the approximation given by the remainder term. □

5.5 Simpson's Rule

Suppose we need a numerical estimation of an integral of the form

$$\int_a^b f(x) \, dx. \tag{5.5}$$

We assume that all derivatives of f formed in the next discussion exist and are continuous. The interval $[a, b]$ is partitioned into $2n$ subintervals, each of length $\Delta x = (b - a)/2n$, and $f(x)$ is approximated by a quadratic polynomial on the first two subintervals, another quadratic polynomial on the third and fourth subintervals, and so on. Of course, polynomials are easy to integrate, and the sum of the integrals of the approximating polynomials is used to approximate (5.5). In order to carry out the integration of the polynomials, we mention Lagrange interpolation. Let $\{x_0, x_1, \ldots, x_n\}$ be $n + 1$ distinct points. Define the function

$$l_i(x) = \prod_{j \ne i} \frac{x - x_j}{x_i - x_j}.$$

Then $l_i(x_i) = 1$ and $l_i(x_j) = 0$ if $j \ne i$. The polynomial

$$p_n(x) = \sum_{i=0}^n f(x_i) l_i(x)$$

interpolates $f(x)$ at $\{x_0, x_1, \ldots, x_n\}$; that is, $p_n(x) = f(x)$ at each x_i. We call $h = \Delta x$. We now restrict our attention to the first two intervals: $x_1 = x_0 + h$, $x_2 = x_0 + 2h$. On $[x_0, x_2]$ we have

$$p_2(x) = f(x_0)l_0(x) + f(x_1)l_1(x) + f(x_2)l_2(x),$$

where

$$l_0(x) = \frac{1}{2h^2}(x - x_0 - h)(x - x_0 - 2h), \qquad l_1(x) = -\frac{1}{h^2}(x - x_0)(x - x_0 - 2h),$$

$$l_2(x) = \frac{1}{2h^2}(x - x_0)(x - x_0 - h).$$

So $p_2(x)$ is a quadratic approximation to f on $[x_0, x_2]$. The integral $\int_{x_0}^{x_2} p_2(x)\,dx$, which we denote by $S_{[x_0,x_2]}$, is after simplification

$$S_{[x_0,x_2]} = \frac{h}{3}\,[f(x_0) + 4f(x_1) + f(x_2)].$$

To find the difference (error) between the integral $\int_{x_0}^{x_2} f(x)\,dx$ and its approximation $S_{[x_0,x_2]}$ we apply Taylor series to f in the integrand and to $S_{[x_0,x_2]}$. Define

$$F(x) = \int_{x_0}^{x} f(t)\,dt.$$

By Theorem 2.5 we have $F'(x) = f(x)$, $F''(x) = f'(x)$, etc. Then

$$F(x_0 + 2h) = F(x_0) + F'(x_0)2h + \cdots + \frac{F^{(5)}(x_0)}{5!}\,(2h)^5 + O(h^6)$$

$$= f(x_0)2h + \cdots + \frac{f^{(4)}(x_0)}{5!}\,(2h)^5 + O(h^6),$$

$$f(x_0 + h) = f(x_0) + f'(x_0)h + \cdots + \frac{f^{(4)}(x_0)}{4!}\,(h)^4 + O(h^5),$$

and

$$f(x_0 + 2h) = f(x_0) + f'(x_0)2h + \cdots + \frac{f^{(4)}(x_0)}{4!}\,(2h)^4 + O(h^5).$$

Since

$$\int_{x_0}^{x_2} f(x)\,dx = F(x_0 + 2h)$$

substituting the Taylor series for the various terms and simplifying, we get

$$\int_{x_0}^{x_2} f(x)\,dx - S_{[x_0,x_2]} = \frac{-h^5}{90}\,f^{(4)}(x_0) + O(h^6).$$

Note that quadratic approximation to f implies the exact cubic approximation in h for the integral. Summing over all pairs of intervals, we get the difference

$$\int_a^b f(x)\,dx - S_{[a,b]} = \frac{-h^5}{90} f^{(4)}(x_0) + \cdots + \frac{-h^5}{90} f^{(4)}(x_{2n-2}) + O(h^6) + \cdots + O(h^6),$$

where $S_{[a,b]}$ is the Simpson approximation given by

$$S_{[a,b]} = S_{[x_0,x_2]} + \cdots + S_{[x_{n-2},x_n]}$$

$$= \frac{h}{3} [f(x_0) + 4f(x_1) + 2f(x_2) + \cdots + 4f(x_{2n-1}) + f(x_{2n})].$$

Let M and m denote the maximum and minimum values, respectively, of $f^{(4)}(x)$ on $[a, b]$. Then

$$nm \le f^{(4)}(x_0) + \cdots + f^{(4)}(x_{2n-2}) \le nM,$$

so that

$$m \le \frac{f^{(4)}(x_0) + \cdots + f^{(4)}(x_{2n-2})}{n} \le M.$$

So by the intermediate value property, for some $\xi \in [a, b]$ we have

$$\frac{f^{(4)}(x_0) + \cdots + f^{(4)}(x_{2n-2})}{n} = f^{(4)}(\xi).$$

Since $b - a = 2nh$, we can write the sum of the n terms

$$\frac{-h^5}{90} f^{(4)}(x_0) + \cdots + \frac{-h^5}{90} f^{(4)}(x_{2n-2}) = \frac{-h^4}{180} f^{(4)}(\xi)(b - a).$$

Similarly, the terms

$$O(h^6) + \cdots + O(h^6) = nO(h^6) = ((b-a)/2h)O(h^6) = O(h^5).$$

The error, $(-h^4/180)f^{(4)}(\xi)(b - a) + O(h^5)$, more simply put, is

$$\int_a^b f(x)\,dx - S_{[a,b]} = O(h^4).$$

This expression for the error is of theoretical interest: if $b - a$ and the higher derivatives of $f(x)$ are not large, then Simpson's rule is very accurate for small h. In practice, we seldom know the fourth derivative of a function, and instead keep doubling the number of partitions until (within a specified tolerance) convergence, noting that the sum of function evaluations at the interior points at one iteration gives the sum of the function evaluations with even indices at the next iteration, hence does not need to be recomputed. For other methods including Rhomberg's method, see Patel [72].

Remark 5.1. Procedures using a specific tolerance and more and more "precise" calculations, such as the doubling practice mentioned above, are widely used in applications, e.g., in the solution of differential equations. The reader should be aware that they do not guarantee convergence of the method, but are merely a sign that we may halt our calculations. We could, for example, try directly calculating partial sums for the series $\sum 1/n$ and, using any tolerance, always get some finite result. Moreover, using standard summation on a computer we even obtain a "precise" value for the series since, as soon as $1/n$ becomes smaller than the least nonzero number in the computer arithmetic, the partial sums will not change. But the series diverges, of course, so the use of such procedures calls for caution.   □

5.6 Taylor's Method

Now we apply Taylor's expansion to the solution of the initial-value problem

$$y' = f(x, y), \quad y(a) = y_0. \tag{5.6}$$

Suppose $y(x)$ is a solution and a numerical estimate of $y(b)$ is to be computed. We suppose $y(x) \in C^{(p+1)}[a, b]$, which means that y has $p + 1$ continuous derivatives on $[a, b]$. Assuming $y(x_n)$ has been computed (accurately) at x_n, we want to compute y at the next value of x, at $x_{n+1} = x_n + h$. By Taylor's theorem

$$y(x_{n+1}) = y(x_n) + y'(x_n)h + y''(x_n)\frac{h^2}{2} + \cdots + y^{(p)}(x_n)\frac{h^p}{p!} + y^{(p+1)}(\xi)\frac{h^{p+1}}{(p+1)!}.$$

Since $y(x)$ is an unknown function of x, we do not know $y'(x), y''(x), \ldots$, explicitly. By (5.6) we do know $y'(x)$ in terms of $x, y(x)$ and by the chain rule we can express $y''(x), y'''(x), \ldots$ in terms of $x, y(x)$. To simplify notation, we write $\partial f/\partial x$ as f_x, $\partial f/\partial y$ as f_y, etc. Since

$$y'(x) = f(x, y(x)), \tag{5.7}$$

by the chain rule (with $y = y(x)$)

$$y''(x) = f'(x, y(x)) = f_x(x, y) + f_y(x, y)y'(x) = f_x(x, y) + f_y(x, y)f(x, y).$$

We shorten the notation for the right-hand side so

$$y''(x) = (f_x + f_y f)\big|_{(x,y)}. \tag{5.8}$$

Similarly,

$$y''' = (f_{xx} + 2f f_{xy} + f_{yy}f^2 + f_x f_y + f_y^2 f)\big|_{(x,y)}. \tag{5.9}$$

Similar but more complicated expressions hold for the higher derivatives. We write Taylor's series for $y(x_n + h)$ as

$$y(x_{n+1}) = y(x_n) + h\Phi_p(x_n, y(x_n)) + y^{(p+1)}(\xi)\frac{h^{p+1}}{(p+1)!},$$

where

$$\Phi_p(x, y) = \left(f + \frac{h}{2}(f_x + f_y f) + \cdots + \frac{h^{(p-1)}}{p!} f^{(p-1)} \right)\Big|_{(x,y)},$$

where the term $f^{(p-1)}(x, y(x))$ is the $(p-1)$th derivative with respect to x and can be expanded as in (5.7), (5.8), etc. Taylor's algorithm of order p uses the Taylor polynomial of degree p to get the next approximation to $y(x_{n+1})$. In other words, let y_n denote the last computed approximation to the exact value $y(x_n)$. Then the next approximation

$$y_{n+1} = y_n + h\Phi_p(x_n, y_n).$$

The special case of $p = 1$ gives the familiar Euler method:

$$y_{n+1} = y_n + hf(x_n, y_n).$$

Of course, the remainder term has been dropped, so at each step there is a (discretization) error caused by approximating the value of y by its Taylor series of order p. To simplify the discussion, we ignore any errors due to roundoff in floating point arithmetic. A local error can grow along with a solution, and so at a later stage earlier discretization errors might have grown along with the solution. To see how the error can grow, call the difference at x_n between the exact solution and the computed approximation $e_n = y(x_n) - y_n$. Then

$$e_{n+1} - e_n = y(x_{n+1}) - y_{n+1} - (y(x_n) - y_n)) = y(x_{n+1}) - y(x_n) - (y_{n+1} - y_n)$$

$$= h\Phi_p(x_n, y(x_n)) + y^{(p+1)}(\xi_n) \frac{h^{p+1}}{(p+1)!} - h\Phi_p(x_n, y_n). \qquad (5.10)$$

We now add the hypothesis that there exists a positive constant L such that for all $u, v \in \mathbb{R}$ and all $x \in [a, b]$,

$$|\Phi_p(x, u) - \Phi_p(x, v)| \le L|u - v|. \qquad (5.11)$$

Since we have assumed $y(x) \in C^{(p+1)}[a, b]$, there is some constant Y such that

$$|y^{(p+1)}(x)| \le Y \qquad (x \in [a, b]). \qquad (5.12)$$

Then by (5.10)–(5.12),

$$|e_{n+1}| \le |e_n| + hL|y(x_n) - y_n| + Y\frac{h^{p+1}}{(p+1)!},$$

hence

$$|e_{n+1}| \le (1 + hL)|e_n| + Y\frac{h^{p+1}}{(p+1)!}. \qquad (5.13)$$

To see how quickly e_n can grow we construct, by replacing inequality with equality, a sequence $\{z_n\}$ that dominates $\{|e_n|\}$. In other words, assuming our initial condition $y(a) = y_0$ is correct, we define $z_0 = e_0 = 0$ and, for all positive n,

$$z_{n+1} = (1 + hL)z_n + Y\frac{h^{p+1}}{(p+1)!} \,.$$

Call

$$B = Yh^{p+1}/(p+1)! \tag{5.14}$$

so

$$z_1 = B,$$

$$z_2 = (1 + hL)z_1 + B = ((1 + hL) + 1)B,$$

$$\vdots$$

$$z_n = ((1 + hL)^{n-1} + \cdots + 1)B. \tag{5.15}$$

Summing the geometric series, we get

$$z_n = B\frac{(1 + hL)^n - 1}{1 + hL - 1} = \frac{(1 + hL)^n - 1}{hL}B\,.$$

By a Taylor series argument $1 + hL < e^{hL}$, so

$$z_n \leq \frac{e^{hLn} - 1}{hL}B\,.$$

Since our x values x_n are in $[a, b]$, $nh \leq (b - a)$, and, using (5.14), we have

$$|e_n| \leq z_n \leq \frac{Y}{L}\left(e^{L(b-a)} - 1\right)\frac{h^p}{(p+1)!} \to 0 \qquad (h \to 0)\,.$$

Because Y and L are constants, at least in theory Taylor's method provides an approximation sequence that converges to the exact solution as $h \to 0$. The error bound, although comforting, is not useful in practice. However, we can compare the results at the same final point b for two different choices of h, say h and $h/2$, and (usually) safely assume that h has been made sufficiently small if the results agree to within a specified tolerance. See [72] for more sophisticated methods of choosing stepsize h appropriately.

The Taylor coefficients can be computed efficiently by *automatic differentiation* [61]. Suppose, for example, we are given the problem

$$y' = 1 - y/x, \quad y(2) = 2\,,$$

and wish to estimate $y(3)$ using Taylor's method of order 5. We do not need to use symbolic manipulation to get the derivative expressions. Instead we can use automatic differentiation to get the derivative values recursively. From the differential equation we find that

$$xy' = x - y,$$
$$xy'' + y' = 1 - y' \text{ so } y'' = (1 - 2y')/x,$$
$$xy''' + y'' = -2y'' \text{ so } y''' = -3y''/x,$$
$$y'''' = -4y'''/x,$$
$$y''''' = -5y''''/x,$$

and so on. Using the given initial condition, we can compute in this way:

$y(2) = 2,$	$y'''(2) = -3/4,$
$y'(2) = 0,$	$y''''(2) = 3/2,$
$y''(2) = 1/2,$	$y'''''(2) = -15/4.$

So we have

$$y(3) = 2 + (1/2)(1/2)(3-2)^2 + +(1/6)(-3/4) + \cdots = 2.15625.$$

The exact solution $y = x/2 + 2/x$ has the value $y(3) = 3/2 + 2/3 = 2.16666\ldots$.

5.7 Special Functions of Mathematical Physics

Applied science is replete with so-called special functions, many of which satisfy interesting inequalities [1, 4, 31, 87].

Example 5.7. If $\mathrm{Re}[z] > 0$, the gamma function $\Gamma(z)$ is given by Euler's integral of the second kind:

$$\Gamma(z) = \int_0^\infty t^{z-1} e^{-t}\, dt.$$

When $z = x$ where x is real and positive, $\Gamma(x)$ can be differentiated any number of times, with

$$\frac{d^n \Gamma(x)}{dx^n} = \int_0^\infty t^{x-1} e^{-t} (\ln t)^n\, dt$$

for any $x > 0$. By the Cauchy–Schwarz inequality,

$$|\Gamma'(x)|^2 \le \int_0^\infty (t^{(x-1)/2} e^{-t/2})^2\, dt \int_0^\infty (t^{(x-1)/2} e^{-t/2} \ln t)^2\, dt$$

and we obtain

$$|\Gamma'(x)|^2 \le \Gamma(x)\Gamma''(x).$$

In the complex-argument case,

$$|\Gamma(x+iy)| = \left| \int_0^\infty t^{x-1} e^{-t} t^{iy} dt \right| \leq \int_0^\infty |t^{x-1} e^{-t}| \, |t^{iy}| \, dt \, ,$$

where $|t^{iy}| = |e^{iy \ln t}| = 1$, and hence

$$|\Gamma(x+iy)| \leq |\Gamma(x)| \, .$$

Of course, for positive integer arguments the gamma function reduces to the factorial function, with $\Gamma(n) = (n-1)!$ (see Problem 5.7).   □

Example 5.8. For $x > 0$ and $n = 0, 1, 2, \ldots$, a sequence of exponential integrals

$$E_n(x) = \int_1^\infty \frac{e^{-xt}}{t^n} \, dt$$

may be defined. The observation

$$\left(\int_1^\infty \frac{e^{-xt}}{t^n} \, dt \right)^2 = \left(\int_1^\infty \frac{e^{-xt/2}}{t^{(n-1)/2}} \frac{e^{-xt/2}}{t^{(n+1)/2}} \, dt \right)^2 \leq \int_1^\infty \left(\frac{e^{-xt/2}}{t^{(n-1)/2}} \right)^2 dt \int_1^\infty \left(\frac{e^{-xt/2}}{t^{(n+1)/2}} \right)^2 dt$$

leads to the inequality

$$E_n^2(x) \leq E_{n-1}(x) E_{n+1}(x) \qquad (n \in \mathbb{N})$$

for the exponential integrals.   □

Example 5.9. The function $T_n(x) = \cos(n \cos^{-1} x)$ for $n \in \mathbb{N}$ is the Chebyshev polynomial of the first kind, order n. Obviously,

$$|T_n(x)| \leq 1 \qquad (-1 \leq x \leq 1) \, .$$

With $x = \cos p$, differentiation gives

$$\frac{dT_n(x)}{dx} = -\frac{1}{\sin p} \frac{dT_n(p)}{dp} = n \frac{\sin np}{\sin p} \, .$$

The maximum occurs at $p = 0$, and we obtain

$$\left| \frac{dT_n(x)}{dx} \right| \leq n^2 \qquad (-1 \leq x \leq 1)$$

for the Chebyshev polynomials.   □

Example 5.10. The Bessel function of the first kind and integer order n may be defined for $-\infty < x < \infty$ by the series expansion

$$J_n(x) = \sum_{m=0}^\infty \frac{(-1)^m (x/2)^{2m+n}}{m! \, (m+n)!} \, ,$$

or by the integral representation

$$J_n(x) = \frac{1}{\pi} \int_0^\pi \cos(nt - x \sin t)\, dt.$$

Immediately from the integral representation

$$|J_n(x)| \le \frac{1}{\pi} \int_0^\pi |\cos(nt - x \sin t)|\, dt \le \frac{1}{\pi} \int_0^\pi (1)\, dt = 1.$$

Other useful properties of $J_n(x)$, such as

$$J_{-n}(x) = (-1)^n J_n(x)$$

may be derived from the series expansion. Still others follow from the generating function relation

$$\exp\left[\frac{x}{2}\left(t - \frac{1}{t}\right)\right] = \sum_{n=-\infty}^{\infty} J_n(x) t^n,$$

among these the symmetry property

$$J_n(-x) = (-1)^n J_n(x),$$

the fact that $J_0(0) = 1$, and the addition theorem

$$J_n(x + y) = \sum_{m=-\infty}^{\infty} J_m(x) J_{n-m}(y).$$

Putting $y = -x$ and $n = 0$ we obtain

$$J_0(0) = 1 = \sum_{m=-\infty}^{\infty} J_m(x) J_{-m}(-x)$$

so that

$$1 = J_0(x)J_0(-x) + \sum_{m=1}^{\infty}[J_{-m}(x)J_m(-x) + J_m(x)J_{-m}(-x)] = J_0^2(x) + 2\sum_{m=1}^{\infty} J_m^2(x).$$

Hence the bound

$$|J_m(x)| \le 1/\sqrt{2} \qquad (m \in \mathbb{N}).$$

It is also of interest to note the *interlacing property* of the zeros of consecutively ordered Bessel functions. We take Rolle's theorem, along with the differentiation formulas

$$[x^n J_n(x)]' = x^n J_{n-1}(x) = -x^{-n} J_{n+1}(x)$$

which hold for any $x > 0$. With $n = k$ and $n = k + 1$, we obtain

$$[x^k J_k(x)]' = -x^{-k} J_{k+1}(x), \qquad [x^{k+1} J_{k+1}(x)]' = x^{k+1} J_k(x),$$

respectively. The first of these equations implies that between any two zeros of J_k, J_{k+1} has at least one zero; but the second implies that between any two zeros of J_{k+1}, J_k has at least one zero. Hence, each function has one and only one zero between each pair of zeros of the other, and the interlacing property is established. □

Example 5.11. The Legendre polynomials $P_n(x)$, for $n = 0, 1, 2, \ldots$, are solutions to a certain ordinary differential equation; they are also given [31] by the Laplace integral formula

$$P_n(x) = \frac{1}{\pi} \int_0^\pi [x + \sqrt{x^2 - 1} \cos t]^n \, dt = \frac{1}{\pi} \int_0^\pi [x + i\sqrt{1 - x^2} \cos t]^n \, dt$$

for $|x| \leq 1$, and possess many other properties, among them the integral

$$\int_{-1}^1 x^m P_n(x) \, dx = 0 \qquad (0 \leq m < n),$$

and the recursion formula

$$(x^2 - 1)P_n'(x) = n[xP_n(x) - P_{n-1}(x)] \qquad (|x| < 1).$$

By the Laplace formula,

$$\pi|P_n(x)| \leq \int_0^\pi |x + i\sqrt{1 - x^2} \cos t|^n \, dt = \int_0^\pi [x^2 + (1 - x^2)\cos^2 t]^{n/2} \, dt$$

$$\leq \int_0^\pi [x^2 + (1 - x^2)]^{n/2} \, dt.$$

Hence the upper bound

$$|P_n(x)| \leq 1 \qquad (|x| \leq 1, \ n = 0, 1, 2, \ldots).$$

Alternatively,

$$\pi|P_n(x)| \leq \int_0^\pi [x^2 + (1 - x^2)\cos^2 t]^{n/2} \, dt = 2\int_0^{\pi/2} [1 - (1 - x^2)\sin^2 t]^{n/2} \, dt$$

$$\leq 2\int_0^{\pi/2} \left[1 - (1 - x^2)\left(\frac{2t}{\pi}\right)^2\right]^{n/2} dt < 2\int_0^{\pi/2} \left(\exp\left[-\frac{4(1 - x^2)t^2}{\pi^2}\right]\right)^{n/2} dt.$$

The last step follows from the inequality

$$e^{-x} > 1 - x \qquad (x > 0).$$

Then

$$\pi|P_n(x)| \leq 2\int_0^{\pi/2} \exp\left[-\frac{2n(1 - x^2)t^2}{\pi^2}\right] dt \leq 2\int_0^\infty \exp\left[-\frac{2n(1 - x^2)t^2}{\pi^2}\right] dt.$$

The remaining integral exists in closed form for $n \neq 0$ and gives us

$$|P_n(x)| \leq \sqrt{\frac{\pi}{2n(1-x^2)}} \qquad (|x| < 1, \; n \in \mathbb{N}).$$

The recursion relation can be treated with the triangle inequality,

$$|P_n'(x)| = \frac{n|xP_n(x) - P_{n-1}(x)|}{|x^2 - 1|} \leq n\frac{|xP_n(x)| + |P_{n-1}(x)|}{|x^2 - 1|} \leq n\frac{|x| + 1}{|x^2 - 1|} = n\frac{|x| + 1}{||x|^2 - 1|},$$

giving the upper bound

$$|P_n'(x)| \leq \frac{n}{1 - |x|} \qquad (|x| < 1)$$

on the first derivatives. $\qquad\square$

We spend a bit more time on the Legendre polynomials.

Example 5.12. The Legendre polynomials are *orthogonal polynomials*. A family of real-valued polynomials $p_n(x)$ for $n = 0, 1, 2, \ldots$ is said to be orthogonal with respect to a weight function $w(x) > 0$ on $[a, b]$ if for $m \neq n$

$$\int_a^b p_n(x)p_m(x)w(x)\,dx = 0. \tag{5.16}$$

Taking functions f that have finite norm

$$\|f\|^2 = \int_a^b f^2(x)w(x)\,dx \tag{5.17}$$

and introducing the corresponding inner product

$$\langle f, g \rangle = \int_a^b f(x)g(x)w(x)\,dx, \tag{5.18}$$

we can apply the Fourier theory with the orthonormal set

$$\frac{p_0(x)}{\|p_0\|}, \frac{p_1(x)}{\|p_1\|}, \frac{p_2(x)}{\|p_2\|}, \ldots$$

The series

$$\sum_{k=0}^{\infty} \left\langle f, \frac{p_k}{\|p_k\|} \right\rangle \frac{p_k(x)}{\|p_k\|}$$

converges relative to the norm (5.17) and Bessel's inequality takes the form

$$\sum_{k=0}^{\infty} \frac{\langle f, p_k \rangle^2}{\|p_k\|^2} \leq \|f\|^2.$$

An interesting fact about orthogonal polynomials is the ease with which a bound can be placed on the locations of their zeros. Putting $m = 0$ in (5.16) we have

$$\int_a^b p_n(x)w(x)\,dx = 0 \qquad (n \geq 1),$$

and it is clear that there is at least one $x \in (a, b)$ at which $p_n(x)$ changes sign. If all such points are denoted by $x_1, \ldots x_k$, then the quantity $p_n(x)(x - x_1)\cdots(x - x_k)w(x)$ never changes sign in $[a, b]$. However, $k < n$ would imply that

$$\int_a^b p_n(x)(x - x_1)\cdots(x - x_k)w(x)\,dx = 0,$$

because $p_n(x)$ is orthogonal to any polynomial of degree less than n. (This follows from the fact that any polynomial of degree less than n can be written uniquely as a linear combination of the polynomials $p_j(x)$ for $j < n$.) So $k \geq n$, hence $k = n$ because $p_n(x)$ cannot have more than n zeros. Our conclusion: the zeros of $p_n(x)$ are all real, distinct, and located in (a, b).   □

Example 5.13. Suppose $P_n(x)$, a polynomial of degree n, is normalized to its leading coefficient c_n to give the related polynomial $\pi_n(x) = P_n(x)/c_n$. It turns out that $\pi_n(x)$ has a smaller norm on $(-1, 1)$ than any other polynomial $f_n(x)$ of degree n with leading coefficient 1. To show this, we define the difference polynomial

$$d_{n-1}(x) = f_n(x) - \pi_n(x)$$

of degree $n - 1$, and note that

$$\|f_n(x)\|^2 = \int_{-1}^1 [d_{n-1}(x) + \pi_n(x)]^2\,dx$$

$$= \int_{-1}^1 d_{n-1}^2(x)\,dx + 2\int_{-1}^1 d_{n-1}(x)\pi_n(x)\,dx + \int_{-1}^1 \pi_n^2(x)\,dx$$

$$= \|d_{n-1}(x)\|^2 + \|\pi_n(x)\|^2.$$

Hence $\|f_n(x)\|^2 \geq \|\pi_n(x)\|^2$.   □

We can sometimes make direct use of power series expansions in establishing bounds for special functions.

Example 5.14. The modified Bessel function of the first kind and zeroth order is given by

$$I_0(x) = \sum_{n=0}^\infty \frac{x^{2n}}{(n!\,2^n)^2}.$$

It is easily shown by induction that $(n!\, 2^n)^2 \geq (2n)!$ for any nonnegative integer n. Hence

$$I_0(x) \leq \sum_{n=0}^{\infty} \frac{x^{2n}}{(2n)!} = \cosh x.$$

See also Problem 5.11.   □

We hasten to point out that bounds for interesting special functions are not always as easy to obtain as those we have seen here. For instance, much more work is needed to bound the Hermite polynomials $H_n(x)$ that are important in quantum mechanics. The reader can see Indritz [38] for a treatment of these functions, including an outline of steps leading to the inequality

$$|H_n(x)| \leq (2^n n!)^{1/2} e^{x^2/2}.$$

5.8 A Projectile Problem

Suppose an object is thrown straight upward with initial speed v_0. If the drag due to air resistance is directly proportional to instantaneous speed, which part of the subsequent motion would take longer: the upward flight, or the return trip?

Newton's second law dictates that the velocity $v_u(t)$ for the upward motion be described by

$$m\frac{dv_u}{dt} = -mg - kv_u,$$

where m is the mass of the object, g is the free-fall acceleration constant, and k is the proportionality constant quantifying air resistance. With the given initial condition, this equation has solution

$$v_u(t) = \gamma g[(1 + \alpha)e^{-t/\gamma} - 1],$$

where $\gamma = m/k$ and $\alpha = v_0/\gamma g$. The time t_u to complete the upward motion can be found from the condition $v_u(t_u) = 0$; it is

$$t_u = \gamma \ln(1 + \alpha)$$

and the maximum height reached is

$$h = \int_0^{t_u} v_u(t)\, dt = \gamma^2 g[\alpha - \ln(1 + \alpha)].$$

Referring a new time origin to the start of the downward motion, the speed v_d for the downward trip is governed by

$$\frac{dv_d}{dt} + \frac{1}{\gamma}v_d = g.$$

When subjected to the initial condition $v_d(0) = 0$, the solution becomes

$$v_d(t) = \gamma g(1 - e^{-t/\gamma}).$$

The time interval t_d for completion of the downward motion must then satisfy

$$\int_0^{t_d} v_d(t)\, dt = h$$

or

$$\gamma g t_d + \gamma^2 g(e^{-t_d/\gamma} - 1) = \gamma^2 g[\alpha - \ln(1 + \alpha)].$$

This is a transcendental equation for the unknown t_d; introducing the variable $T_d = t_d/\gamma$, we can write it as $F(T_d) = 0$ where

$$F(x) = x + e^{-x} - (1 + \alpha) + \ln(1 + \alpha).$$

Because $F'(x) = 1 - e^{-x} > 0$, we know that $F(x)$ is strictly increasing. Defining $T_u = t_u/\gamma$ we have

$$F(T_u) = 2\ln(1 + \alpha) - (1 + \alpha) + \frac{1}{1 + \alpha} = 2\ln(1 + \alpha) - \alpha\left(\frac{2 + \alpha}{1 + \alpha}\right) < 0,$$

where the last inequality is easily verified by differentiation. This and the monotonicity of F are enough to conclude (Fig. 5.1) that $T_u < T_d$. Hence, the object spends more time in its descent than it does in its ascent.

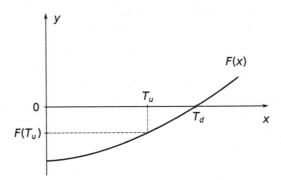

Fig. 5.1 Times for projectile motion. The function F, by its definition, passes through zero at $x = T_d$. Since $F(T_u) < 0$ and F is increasing, we have $T_u < T_d$

Of course, the physical reliability of this conclusion depends on the correctness of the model employed. It is well known that in many (if not most) situations, drag is proportional to the *square* of the speed—see Glaister [30] for a treatment of this case. For a more general analysis with arbitrary air resistance, see de Alwis [17].

5.9 Geometric Shapes

It is worthwhile to examine a few applications of inequalities to simple geometrical objects.

Example 5.15. A polyhedron is a solid figure bounded by planes. Such a figure can be considered as a union of a finite number of polygonal faces. The faces are joined along line segments called edges, and at the two ends of each edge are points called vertices. Among the most beautiful of the polyhedra are the regular polyhedra, where the faces are all congruent regular polygons. That there exist only five of these—the *Platonic solids*—can be shown by simple arguments with inequalities. The proof is based on Euler's formula, which states that for any simple polyhedron, the number of faces F, the number of edges E, and the number of vertices V are related by the equation

$$F - E + V = 2. \tag{5.19}$$

For instance, the cube has $F = 6$, $E = 12$, $V = 8$, while the tetrahedron has $F = 4$, $E = 6$, $V = 4$. A precise definition of the term "simple" would require a topological digression that would send us too far afield; suffice it to say that we rule out shapes with holes in them, such as toroidal-shaped polyhedra [5].

Consider then a simple, regular polyhedron. Because all the face polygons are identical, we may define a constant σ as the number of edges per face, and another constant v as the number of edges meeting at each vertex. Clearly $\sigma \geq 3$ and $v \geq 3$. Moreover, $2E = \sigma F = vV$, as each edge has two vertices and is shared by two faces. Elimination of F and V from (5.19) gives

$$1/\sigma + 1/v = 1/2 + 1/E. \tag{5.20}$$

Now because

$$1/\sigma + 1/v > 1/2,$$

we must rule out the possibility that both $\sigma > 3$ and $v > 3$. Putting $\sigma = 3$ in (5.20) we get

$$1/v - 1/6 = 1/E > 0,$$

and hence the restriction $3 \leq v \leq 5$; putting instead $v = 3$ we get $3 \leq \sigma \leq 5$. The five permissible combinations can be tabulated as follows:

σ	v	E	$F = 2E/\sigma$	$V = 2E/v$	Object
3	3	6	4	4	Tetrahedron
3	4	12	8	6	Octahedron
3	5	30	20	12	Icosahedron
4	3	12	6	8	Cube
5	3	30	12	20	Dodecahedron

These are the Platonic solids.   □

The *isoperimetric inequalities* provide information regarding the extremal properties of geometric shapes. We encounter one of these inequalities in our next example (see [55]).

Example 5.16. Let $f(x)$ be periodic with period L. We showed previously that under suitable restrictions it can be represented by a uniformly convergent Fourier series. The form of the series in this case is

$$f(x) = a_0 + \sum_{n=1}^{\infty} a_n \cos \frac{2\pi n}{L} x + b_n \sin \frac{2\pi n}{L} x.$$

Similarly, let us take another function $F(x)$ of the same period and having a uniformly convergent Fourier series. Then

$$F(x) = A_0 + \sum_{n=1}^{\infty} A_n \cos \frac{2\pi n}{L} x + B_n \sin \frac{2\pi n}{L} x.$$

Integration of the product of these two functions over $[0, L]$ gives Parseval's identity

$$\int_0^L f(x)F(x)\, dx = \frac{L}{2}\left[2a_0 A_0 + \sum_{n=1}^{\infty} (a_n A_n + b_n B_n) \right]. \qquad (5.21)$$

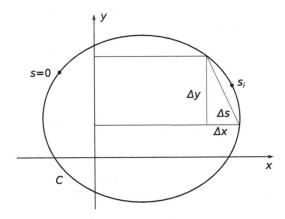

Fig. 5.2 Derivation of an isoperimetric inequality

Consider a simple, smooth, closed plane curve C with known length L as in Fig. 5.2. We ask what *shape* C must have in order that its enclosed area A be maximized. (This is an extremal problem, but not of the type normally encountered in calculus courses.) We choose a reference point P on C, and define a parameter s to measure arc length along C to the point (x, y) as shown. We assume that $x(s)$ and $y(s)$, each with period L, are sufficiently smooth (i.e., differentiable) so that we may represent them as Fourier series:

$$x(s) = a_0 + \sum_{n=1}^{\infty} a_n \cos \frac{2\pi n}{L} s + b_n \sin \frac{2\pi n}{L} s,$$

$$y(s) = A_0 + \sum_{n=1}^{\infty} A_n \cos \frac{2\pi n}{L} s + B_n \sin \frac{2\pi n}{L} s.$$

Moreover, uniform convergence of the following formal termwise derivatives of the series permits the termwise differentiation indicated:

$$x'(s) = \sum_{n=1}^{\infty} \frac{2\pi n}{L} \left(b_n \cos \frac{2\pi n}{L} s - a_n \sin \frac{2\pi n}{L} s \right),$$

$$y'(s) = \sum_{n=1}^{\infty} \frac{2\pi n}{L} \left(B_n \cos \frac{2\pi n}{L} s - A_n \sin \frac{2\pi n}{L} s \right).$$

The application of (5.21) to these series and addition of the results gives

$$\int_0^L [x'^2(s) + y'^2(s)] \, ds = \frac{2\pi^2}{L} \sum_{n=1}^{\infty} n^2 (a_n^2 + b_n^2 + A_n^2 + B_n^2).$$

But $x'^2(s) + y'^2(s) \equiv 1$ so that

$$\sum_{n=1}^{\infty} n^2 (a_n^2 + b_n^2 + A_n^2 + B_n^2) = \frac{L^2}{2\pi^2}.$$

Referring again to the figure, we have for the enclosed area

$$A \approx \sum_i x(s_i) \, \Delta y = \sum_i x(s_i) \frac{\Delta y}{\Delta s} \, \Delta s$$

or in the limit

$$A = \int_0^L x(s) y'(s) \, ds = \pi \sum_{n=1}^{\infty} n(a_n B_n - A_n b_n)$$

by (5.21) and use of the differentiated series above. Then

$$L^2 - 4\pi A = 2\pi^2 \sum_{n=1}^{\infty} n^2 (a_n^2 + b_n^2 + A_n^2 + B_n^2) - 4\pi^2 \sum_{n=1}^{\infty} n(a_n B_n - A_n b_n)$$

$$= 2\pi^2 \sum_{n=1}^{\infty} [(na_n - B_n)^2 + (nA_n + b_n)^2 + (n^2 - 1)(b_n^2 + B_n^2)]$$

or, since the right member is nonnegative,

$$A \leq L^2 / 4\pi.$$

This isoperimetric inequality will answer our maximization question. It is apparent that equality holds if and only if: (1) all the Fourier coefficients vanish whenever $n \geq 2$; and (2) $a_1 = B_1$ and $A_1 = -b_1$. Under these conditions

$$x(s) = a_0 + a_1 \cos \frac{2\pi}{L} s + b_1 \sin \frac{2\pi}{L} s,$$

$$y(s) = A_0 - b_1 \cos \frac{2\pi}{L} s + a_1 \sin \frac{2\pi}{L} s.$$

Squaring and adding to eliminate s, we obtain

$$(x - a_0)^2 + (y - A_0)^2 = a_1^2 + b_1^2.$$

Hence of all closed curves of a given length, a circle encloses the greatest area. □

Example 5.17. We can show that of all triangles having a given fixed perimeter, an equilateral triangle encloses the greatest area. The area A of a triangle is given by Heron's formula

$$A = [s(s - a)(s - b)(s - c)]^{1/2},$$

where a, b, c are the side lengths and s is one-half the perimeter p. By the AM–GM inequality,

$$\left(\frac{A^2}{s}\right)^{1/3} = [(s - a)(s - b)(s - c)]^{1/3} \leq \frac{(s - a) + (s - b) + (s - c)}{3} = \frac{s}{3}.$$

Hence

$$A \leq \frac{s^2}{3\sqrt{3}} = \frac{p^2}{12\sqrt{3}}.$$

Equality is attained only if the numbers $s - a$, $s - b$, $s - c$ are all equal. □

5.10 Electrostatic Fields and Capacitance

Electrostatics is the study of stationary electric charges and their mutual effects. Utmost in electrostatics is the conservative nature of the force field, which permits the vector electric intensity to be expressed as the gradient of a potential function. The electrostatic potential $\Phi(x, y, z)$ is produced by electric charge and satisfies Poisson's equation

$$\nabla^2 \Phi = \frac{\partial^2 \Phi}{\partial x^2} + \frac{\partial^2 \Phi}{\partial y^2} + \frac{\partial^2 \Phi}{\partial z^2} = -\frac{\rho}{\epsilon_0},$$

where $\rho = \rho(x, y, z)$ is the density of electric charge (Coulombs/meter^3) and ϵ_0 is the free-space permittivity (a positive constant). The electrostatic potential Φ is convenient because of its scalar nature, and we can study some of its most fundamental properties through basic work with inequalities.

Example 5.18. Consider electric charge distributed with density $\rho(x, y, z)$ throughout a volume region V in unbounded free space. The resulting potential at points (x, y, z) external to V is given by

$$\Phi(x, y, z) = \frac{1}{4\pi\epsilon_0} \int_V \frac{\rho(x', y', z')}{R} \, dx' dy' dz',$$

where R is the distance from (x, y, z) to an element of electric charge at the location (x', y', z') of the differential volume $dx' dy' dz'$. We study the behavior of Φ at large distance from V. For simplicity we assume $\rho(x, y, z) \geq 0$ in V; the following argument is easily modified if negative charge is present. Let the maximum and minimum values of R for a fixed point (x, y, z) be R_M and R_m, respectively. Then

$$\frac{1}{R_M} \leq \frac{1}{R} \leq \frac{1}{R_m}$$

and we obtain

$$\int_V \frac{\rho}{R_M} \, dx' dy' dz' \leq \int_V \frac{\rho}{R} \, dx' dy' dz' \leq \int_V \frac{\rho}{R_m} \, dx' dy' dz'$$

or

$$\frac{Q}{4\pi\epsilon_0} \frac{R}{R_M} \leq R\Phi \leq \frac{Q}{4\pi\epsilon_0} \frac{R}{R_m},$$

where

$$Q = \int_V \rho \, dx' dy' dz'$$

is the total charge in V. As $R \to \infty$, R/R_M and R/R_m both approach 1 and the middle term $R\Phi$ is squeezed to $Q/4\pi\epsilon_0$. Hence $\Phi = O(R^{-1})$ and the potential is said to be *regular at infinity*. $\square$

Example 5.19 (See [90]). Consider a two-dimensional situation where

$$\frac{\partial^2 \Phi}{\partial x^2} + \frac{\partial^2 \Phi}{\partial y^2} = -\frac{\rho(x, y)}{\epsilon_0} \tag{5.22}$$

holds within a bounded domain D of the xy-plane. Let the boundary curve of D be C. We investigate a property possessed by any continuous solution $\Phi(x, y)$ under the condition that ρ is strictly negative so that the forcing function for (5.22) is strictly positive. If Φ is continuous on $D \cup C$, then it must attain a maximum on $D \cup C$, at point $p_0 = (x_0, y_0)$ say. Now $p_0 \in D$ implies that simultaneously

$$\left. \frac{\partial^2 \Phi}{\partial x^2} \right|_{p_0} \leq 0, \qquad \left. \frac{\partial^2 \Phi}{\partial y^2} \right|_{p_0} \leq 0,$$

a contradiction, and hence $p_0 \in C$. Under the given assumptions, Φ must attain its maximum on the boundary contour C.

Next, suppose ρ is merely nonpositive in D. We take B to be an upper bound for Φ on C, and let

$$\Phi(x, y) = \phi(x, y) - \varepsilon(x^2 + y^2),$$

where ϕ is a new function and $\varepsilon > 0$ is arbitrary. Substitution into (5.22) gives

$$\frac{\partial^2 \phi}{\partial x^2} + \frac{\partial^2 \phi}{\partial y^2} = -\frac{\rho(x, y)}{\epsilon_0} + 4\varepsilon.$$

Because ϕ satisfies Poisson's equation with a strictly positive forcing function, we know that ϕ attains its maximum on C. Then, as $\Phi(x, y) \le \phi(x, y)$, we have

$$\max_{(x,y)\in D} \Phi \le \max_{(x,y)\in D} \phi \le \max_{(x,y)\in C} \phi = \max_{(x,y)\in C} [\Phi + \varepsilon(x^2 + y^2)] \le B + \varepsilon \max_{(x,y)\in C} (x^2 + y^2)$$

for every $\varepsilon > 0$. Hence $\Phi(x, y) \le B$ for all $(x, y) \in D$.

The two results obtained above, called *maximum principles*, yield prior knowledge about the behavior of all possible solutions of Poisson's equation and hence of the electric potential under certain prescribed conditions. The case $\rho(x, y) \equiv 0$ of (5.22) is the important Laplace equation

$$\frac{\partial^2 \Phi}{\partial x^2} + \frac{\partial^2 \Phi}{\partial y^2} = 0. \tag{5.23}$$

We suppose for (5.23) that there exist positive numbers b and B such that for every $(x, y) \in C$,

$$b \le \Phi(x, y) \le B. \tag{5.24}$$

Then $\Phi(x, y) \le B$ in D by the maximum principle above. Moreover, $-\Phi$ also satisfies (5.23) and the condition $-B \le -\Phi(x, y) \le -b$ on C. Hence on D we have $-\Phi(x, y) \le -b$ and it follows that (5.24) holds there. In other words, both the maximum and minimum values of any solution to (5.23) in a bounded domain must occur on the boundary of the domain. The reader interested in pursuing this area further could obtain Protter [75]. $\square$

A quantity of interest in electrostatics is the capacitance of a metallic body. Consider a conducting solid with boundary surface A and held at potential Φ_0, and let Φ be the potential produced by the charge on the body. The capacitance of the body is defined as the ratio of the total charge it carries to the potential Φ_0, and is given by

$$C = \frac{\epsilon_0}{\Phi_0^2} \int_{V_e} |\nabla \Phi|^2 \, dV, \tag{5.25}$$

where volume integration is done over the space V_e exterior to A. Based on this relation we can derive an inequality that provides a convenient upper bound on the capacitance. We introduce two new scalar fields f and δ (not having any particular physical interpretation) such that

$$f(x, y, z) = \Phi(x, y, z) + \delta(x, y, z),$$

where $\delta = 0$ on the body A and $\delta \to 0$ at large distance from A so that f satisfies the same boundary conditions as Φ. Notice that

$$\int_{V_e} |\nabla f|^2 \, dV = \int_{V_e} |\nabla \Phi + \nabla \delta|^2 \, dV$$

$$= \int_{V_e} |\nabla \Phi|^2 \, dV + 2 \int_{V_e} \nabla \delta \cdot \nabla \Phi \, dV + \int_{V_e} |\nabla \delta|^2 \, dV,$$

so that by (5.25), Green's formula

$$\oint_S \phi \frac{\partial \psi}{\partial n} \, dS = \int_V \nabla \phi \cdot \nabla \psi \, dV + \int_V \phi \nabla^2 \psi \, dV \tag{5.26}$$

and Laplace's equation, we get

$$\int_{V_e} |\nabla f|^2 \, dV = \frac{\Phi_0^2 C}{\epsilon_0} + 2 \left[\oint_S \delta \frac{\partial \Phi}{\partial n} \, dS - \int_{V_e} \delta \nabla^2 \Phi \, dV \right] + \int_{V_e} |\nabla \delta|^2 \, dV$$

$$= \frac{\Phi_0^2 C}{\epsilon_0} + \int_{V_e} |\nabla \delta|^2 \, dV. \tag{5.27}$$

Because the rightmost term is nonnegative, we have *Dirichlet's principle*

$$C \leq \frac{\epsilon_0}{\Phi_0^2} \int_{V_e} |\nabla f|^2 \, dV. \tag{5.28}$$

Equality is attained when $f = \Phi$, i.e., the actual potential; any other function that fits the same boundary conditions will overestimate C.

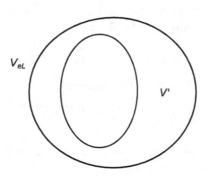

Fig. 5.3 Bound on capacitance

Example 5.20. The capacitance C of a body is less than that of any other body that can completely surround it (Fig. 5.3). Let Φ_L be the actual potential due to the larger body, and take $f = \Phi_L$ at points in the region V_{eL} outside the larger body while taking $f = \Phi_0$ everywhere inside. Then f satisfies the boundary conditions for both bodies and

$$C \leq \frac{\epsilon_0}{\Phi_0^2} \int_{V_e} |\nabla f|^2 \, dV = \frac{\epsilon_0}{\Phi_0^2} \int_{V'} |\nabla \Phi_0|^2 \, dV + \frac{\epsilon_0}{\Phi_0^2} \int_{V_{eL}} |\nabla \Phi_L|^2 \, dV = C_L.$$

For instance, because the capacitance of a sphere is given by an elementary formula, we could get a rough estimate of the capacitance of a cube by inscribing and circumscribing appropriate spheres.   □

The reader interested in more sophisticated uses of Dirichlet's principle is referred to Polya and Szegö [74] who, for instance, show that the capacitance of a convex but otherwise arbitrarily shaped body cannot exceed the capacitance of a certain related prolate spheroid. The capacitance of a prolate spheroid of eccentricity e is well known [85] and is given by

$$C = \frac{8\pi\epsilon_0 ae}{\ln[(1 + e)/(1 - e)]},\tag{5.29}$$

where a is the major semiaxis of the generating ellipse. The capacitance of a convex body can never exceed the capacitance of a prolate spheroid, the major and minor semiaxes of which are the *mean radius* and *surface radius*, respectively, of the body. These last terms are further defined in the reference, where this elegant result is used to attack the difficult problem of better estimating the capacitance of a cube.

Another approach to the estimation of electrical capacitance is based on a geometrical concept of symmetrization. *Steiner symmetrization* of a given solid B with respect to a plane P is an operation which changes B into new solid B' such that:

(a) B' is symmetric with respect to P;
(b) if L is a straight line perpendicular to P, then L intersects B if and only if L intersects B', and both intersections have the same length;
(c) $L \cap B'$ is just one line segment, bisected by P (or, is a point of P in a degenerate case).

P is known as the plane of symmetrization for the operation. For instance, suppose our original solid is the hemispherical ball

$$B = \{(x, y, z): 0 < z < \sqrt{a^2 - x^2 - y^2}\}$$

and P is the $z = 0$ plane. We see that the new solid

$$B' = \left\{(x, y, z): |z| < \frac{1}{2} \sqrt{a^2 - x^2 - y^2}\right\}$$

satisfies the three conditions of the definition of symmetrization; hence, B' is the ellipsoid

$$\frac{x^2}{a^2} + \frac{y^2}{a^2} + \frac{z^2}{(a/2)^2} = 1.$$

Symmetrization has useful properties. First, the solids B and B' have equal volumes (as is easily verified for the example above). Second, the operation does not increase surface area; that is, supposing B has boundary area S and B' has boundary area S', then $S' \leq S$. A similar relation holds between the capacitances C and C' of metallic objects formed in the shapes of B and B', respectively, i.e., that

$$C' \leq C.$$

Polya and Szegö discuss an ingenious application of this idea to the calculation of capacitance of arbitrarily shaped bodies. The main ideas are as follows. We begin with an arbitrarily shaped body $B^{(0)}$ having known volume V and unknown capacitance $C^{(0)}$. Now imagine symmetrizing the body repeatedly, with respect to any number of different successive planes. After the nth such symmetrization, we get a body $B^{(n)}$ whose volume is still V but whose capacitance $C^{(n)}$ satisfies

$$C^{(n)} \leq C^{(n-1)} \leq \cdots \leq C^{(0)}.$$

As $n \to \infty$ we should arrive at a sphere of volume V; letting $e \to 0$ in (5.29), the capacitance of this simple object is $4\pi\epsilon_0 a$ where a is the radius. Since the volume is $V = 4\pi a^3/3$, we can eliminate a from the capacitance expression and assert that

$$C^{(0)} \geq 4\pi\epsilon_0 \sqrt[3]{\frac{3V}{4\pi}}$$

for the body $B^{(0)}$. For a metallic cube of edge length L, for instance, this would yield

$$C \geq 4\pi\epsilon_0 \sqrt[3]{\frac{3L^3}{4\pi}} \approx 7.796\epsilon_0 L$$

as a lower bound.

5.11 Applications to Matrices

Matrix theory and linear algebra contain many references to inequalities. Given an $n \times n$ square matrix of complex elements

$$A = \begin{pmatrix} a_{11} & \cdots & a_{1n} \\ \vdots & \ddots & \vdots \\ a_{n1} & \cdots & a_{nn} \end{pmatrix},$$

an important related matrix is the conjugate transpose $A^\dagger$ of A:

$$A^\dagger = \begin{pmatrix} \bar{a}_{11} & \cdots & \bar{a}_{n1} \\ \vdots & \ddots & \vdots \\ \bar{a}_{1n} & \cdots & \bar{a}_{nn} \end{pmatrix}.$$

In terms of inner products,

$$\langle \mathbf{x}, A\mathbf{y} \rangle = \langle A^\dagger \mathbf{x}, \mathbf{y} \rangle \qquad (\mathbf{x}, \mathbf{y} \in \mathbb{C}^n).$$

The matrix A is called Hermitian or self-adjoint if

$$A = A^\dagger.$$

Because $\langle \mathbf{x}, A\mathbf{y} \rangle = \langle A^\dagger \mathbf{x}, \mathbf{y} \rangle$ for any vectors $\mathbf{x}, \mathbf{y} \in \mathbb{C}^n$ and any complex matrix A, the Hermitian matrix A satisfies

$$\langle \mathbf{x}, A\mathbf{y} \rangle = \langle A\mathbf{x}, \mathbf{y} \rangle. \tag{5.30}$$

Next, we derive an important inequality involving the eigenvalues of a square matrix A. Recall that the eigenvalues $\lambda_1, \ldots, \lambda_n$ of A are scalars satisfying

$$A\mathbf{x} = \lambda_i \mathbf{x}$$

for some nonzero column vectors $\mathbf{x}$ (the corresponding eigenvectors). We first note that if λ is an eigenvalue of the Hermitian matrix A, then λ must be real. To see this, let λ be an eigenvalue with corresponding eigenvector $\mathbf{x}$. Then

$$\langle \mathbf{x}, \mathbf{x} \rangle \overline{\lambda} = \langle \mathbf{x}, \lambda \mathbf{x} \rangle = \langle \mathbf{x}, A\mathbf{x} \rangle = \langle A\mathbf{x}, \mathbf{x} \rangle = \langle \lambda \mathbf{x}, \mathbf{x} \rangle = \lambda \langle \mathbf{x}, \mathbf{x} \rangle.$$

Since $\langle \mathbf{x}, \mathbf{x} \rangle \neq 0$, we have $\overline{\lambda} = \lambda$. In case A is a real matrix, Hermitian means symmetric: $a_{ij} = a_{ji}$ for all i, j. We now restrict our discussion to symmetric matrices. Now suppose λ_1 and λ_2 are two (distinct) eigenvalues of the symmetric matrix A with corresponding eigenvectors $\mathbf{x}_1, \mathbf{x}_2$. Then $\mathbf{x}_1$ and $\mathbf{x}_2$ are orthogonal, i.e.,

$$\langle \mathbf{x}_1, \mathbf{x}_2 \rangle = 0.$$

To see this, we write

$$\langle \mathbf{x}_1, \mathbf{x}_2 \rangle \lambda_2 = \langle \mathbf{x}_1, \lambda_2 \mathbf{x}_2 \rangle = \langle \mathbf{x}_1, A\mathbf{x}_2 \rangle = \langle A\mathbf{x}_1, \mathbf{x}_2 \rangle = \langle \lambda_1 \mathbf{x}_1, \mathbf{x}_2 \rangle = \lambda_1 \langle \mathbf{x}_1, \mathbf{x}_2 \rangle.$$

But $\lambda_1 \neq \lambda_2$ and the result follows. Moreover, we obtain the *generalized orthogonality* property

$$\langle A\mathbf{x}_1, \mathbf{x}_2 \rangle = 0.$$

An important result for Hermitian matrices is that for any Hermitian matrix A there is an orthonormal set of eigenvectors that constitute a basis of $\mathbb{C}^n$. The following several theorems are also useful.

Theorem 5.1. *Let A be an $n \times n$ symmetric (real) matrix. Suppose the (real) eigenvalues $\{\lambda_i\}$ satisfy $\lambda_1 < \lambda_2 < \cdots < \lambda_n$. Define the quadratic form for $\mathbf{x} \in \mathbb{R}^n$ by $Q(\mathbf{x}) = \langle \mathbf{x}, A\mathbf{x} \rangle$. Then, for any $\mathbf{x} \in \mathbb{R}^n$,*

$$\lambda_1 \|\mathbf{x}\|^2 \leq Q(\mathbf{x}) \leq \lambda_n \|\mathbf{x}\|^2.$$

Proof. Let $\{\mathbf{x}_1, \ldots, \mathbf{x}_n\}$ be corresponding eigenvectors. We may assume that each $\mathbf{x}_i$ satisfies $\|\mathbf{x}_i\| = 1$. (Otherwise replace $\mathbf{x}_i$ by $\mathbf{x}_i / \|\mathbf{x}_i\|$.) By our previous observation, the vectors $\{\mathbf{x}_1, \ldots, \mathbf{x}_n\}$ are an orthonormal set. Hence they are linearly independent and form a basis. Thus there exist coefficients $\{c_i\}$ such that

$$\mathbf{x} = \sum_{i=1}^{n} c_i \mathbf{x}_i.$$

Then

$$Q(\mathbf{x}) = \left\langle \sum_{i=1}^{n} c_i \mathbf{x}_i, A \sum_{j=1}^{n} c_j \mathbf{x}_j \right\rangle = \left\langle \sum_{i=1}^{n} c_i \mathbf{x}_i, \sum_{j=1}^{n} c_j \lambda_j \mathbf{x}_j \right\rangle = \sum_{i=1}^{n} c_i^2 \lambda_i$$

so that

$$Q(\mathbf{x}) \le \lambda_n \sum_{i=1}^{n} c_i^2 = \lambda_n \|\mathbf{x}\|^2.$$

Similarly, $\lambda_1 \|\mathbf{x}\|^2 \le Q(\mathbf{x})$. $\qquad\square$

Theorem 5.2 (Sylvester's Criterion). *Let A be a symmetric $n \times n$ real matrix. The quadratic form defined by $Q(\mathbf{x}) = \mathbf{x}^T A \mathbf{x} = \langle \mathbf{x}, A\mathbf{x} \rangle$ for $\mathbf{x} \in \mathbb{R}^n$ is positive definite if and only if the following determinants are all positive:*

$$|a_{11}|, \quad \begin{vmatrix} a_{11} & a_{12} \\ a_{21} & a_{22} \end{vmatrix}, \quad \dots, \quad \begin{vmatrix} a_{11} & \cdots & a_{1n} \\ \vdots & \ddots & \vdots \\ a_{n1} & \cdots & a_{nn} \end{vmatrix}.$$

Proof. Recall from linear algebra that an $n \times n$ symmetric matrix A is called positive definite if $\mathbf{x}^T A \mathbf{x} > 0$ whenever the vector $\mathbf{x} \ne \mathbf{0}$. We discuss the theorem for $n = 2$ and refer the reader to Gelfand [28] for the general case. Suppose $Q(\mathbf{x})$ is positive definite. That is, $Q(\mathbf{x}) > 0$ if $\mathbf{x} \ne \mathbf{0}$. Choose $\mathbf{x} = (1, 0)^T$. Then $Q(\mathbf{x}) = a_{11} > 0$. Now choose $\mathbf{x} = (x_1, 1)^T$. Because $\mathbf{x} \ne \mathbf{0}$ we have $Q(\mathbf{x}) = a_{11} x_1^2 + 2 a_{12} x_1 + a_{22} > 0$ for all x_1 so by our discussion of quadratic inequalities in Chap. 1,

$$\Delta = \begin{vmatrix} a_{11} & a_{12} \\ a_{12} & a_{22} \end{vmatrix} > 0.$$

The converse is proved in a similar way. $\qquad\square$

Theorem 5.3 (Second Derivative Test for n Variables). *Let U be an open set in $\mathbb{R}^n$. Let $f(\mathbf{x}) \in C^{(2)}(U)$. Let $\mathbf{x}_0 \in U$ and suppose $f'(\mathbf{x}_0) = 0$ and $f''(\mathbf{x}_0)$ is positive definite. Then $f(\mathbf{x})$ has a local minimum at $\mathbf{x}_0$. That is, there exists $\delta > 0$ such that $f(\mathbf{x}) > f(\mathbf{x}_0)$ whenever $0 < \|\mathbf{x} - \mathbf{x}_0\| < \delta$.*

Proof. We first discuss some of the terms used. $f(\mathbf{x}) \in C^{(2)}(U)$ means all first and second partial derivatives of f with respect to $\{x_1, \dots, x_n\}$ exist and are continuous in U. $f'(\mathbf{x}_0)$ represents the n-tuple, or row vector $(\partial f(\mathbf{x}_0)/\partial x_1, \dots, \partial f(\mathbf{x}_0)/\partial x_n)$. The second derivative $f''(\mathbf{x}_0)$, or Hessian, is the $n \times n$ matrix whose ijth entry is $\partial^2 f(\mathbf{x}_0)/\partial x_i \partial x_j$. By Sylvester's criterion, the quadratic form $\mathbf{x}^T f''(\mathbf{x}_0)\mathbf{x} = \langle \mathbf{x}, f''(\mathbf{x}_0)\mathbf{x} \rangle$ and hence the second derivative $f''(\mathbf{x}_0)$ is positive definite if and only if the determinants

$$\left| \frac{\partial^2 f}{\partial x_1^2} \right|, \quad \begin{vmatrix} \dfrac{\partial^2 f}{\partial x_1^2} & \dfrac{\partial^2 f}{\partial x_1 \partial x_2} \\[2ex] \dfrac{\partial^2 f}{\partial x_1 \partial x_2} & \dfrac{\partial^2 f}{\partial x_2^2} \end{vmatrix}, \quad \dots, \quad \begin{vmatrix} \dfrac{\partial^2 f}{\partial x_1^2} & \cdots & \dfrac{\partial^2 f}{\partial x_1 \partial x_n} \\[2ex] \vdots & \ddots & \vdots \\[2ex] \dfrac{\partial^2 f}{\partial x_n \partial x_1} & \cdots & \dfrac{\partial^2 f}{\partial x_n^2} \end{vmatrix}$$

are all positive at $\mathbf{x}_0$. By persistence of sign applied to n variables, if the determinants are positive at $\mathbf{x}_0$, then they are positive nearby. Choose $\delta > 0$ such that $\mathbf{x} \in U$ and all the determinants are positive at $\mathbf{x}$ whenever $\|\mathbf{x} - \mathbf{x}_0\| < \delta$. Now let $0 < \|\Delta\mathbf{x}\| < \delta$. The following generalizes Eq. (2.13) to n variables:

$$f(\mathbf{x}_0 + \Delta\mathbf{x}) = f(\mathbf{x}_0) + f'(\mathbf{x}_0)\Delta\mathbf{x} + \tfrac{1}{2}(\Delta\mathbf{x})^T f''(\xi)(\Delta\mathbf{x})$$

for some ξ belonging to the line segment $S = \{\mathbf{z}: \mathbf{z} = t\mathbf{x} + (t-1)(\mathbf{x}+\Delta\mathbf{x}), \ 0 \le t \le 1\}$ (see [18, 45]). Because $f'(\mathbf{x}_0) = 0$ and $f''(\xi)$ is positive definite, the result follows by inspection. $\qquad\square$

Another useful concept is the trace of a square matrix M, denoted by tr$[M]$ and defined as the sum of the diagonal elements of M. With $B = A^\dagger A$ we have $b_{ij} = \sum_{k=1}^n \overline{a}_{ki}a_{kj}$ and the trace of B is

$$\sum_{m=1}^n b_{mm} = \sum_{m=1}^n \sum_{k=1}^n \overline{a}_{km}a_{km} = \sum_{m=1}^n \sum_{k=1}^n |a_{km}|^2 \ge 0.$$

Hence

$$\text{tr}[A^\dagger A] \ge 0$$

with equality if and only if A is the zero matrix. We use this simple result to derive another inequality involving matrix eigenvalues. Because

$$\lambda\mathbf{x} - A\mathbf{x} = \lambda I\mathbf{x} - A\mathbf{x} = (\lambda I - A)\mathbf{x}$$

where I is the identity matrix having the same dimension as A, the eigenvalues are conveniently computed as solutions of the characteristic equation

$$\det(\lambda I - A) = 0.$$

For any square matrix S there is a unitary matrix U (i.e., one having $U^{-1} = U^\dagger$) such that $U^\dagger S U$ is upper triangular. This upper triangular matrix $T = U^\dagger S U$ is said to be unitarily similar to S. Since

$$\det(\lambda I - T) = \det(\lambda U^\dagger I U - U^\dagger S U) = \det(U^{-1})\det(\lambda I - S)\det(U) = \det(\lambda I - S)$$

it is apparent that T has the same eigenvalues as does S; moreover, the eigenvalues of T reside along the main diagonal. We use these facts as follows. Suppose A is square with eigenvalues $\lambda_1, \ldots, \lambda_n$. Then $B = U^\dagger A U$ is upper triangular and

$$BB^\dagger = (U^\dagger A U)(U^\dagger A U)^\dagger = (U^\dagger A U)(U^\dagger(U^\dagger A)^\dagger) = (U^\dagger A U)(U^\dagger A^\dagger U) = U^\dagger A A^\dagger U$$

so that $BB^\dagger$ is unitarily similar to $AA^\dagger$. Because the trace of a matrix is the sum of its eigenvalues, we have tr$[BB^\dagger]$ = tr$[AA^\dagger]$ or

$$\sum_{i=1}^n \sum_{j=1}^n |a_{ij}|^2 = \sum_{i=1}^n \sum_{j=1}^n |b_{ij}|^2.$$

But remembering that B is upper triangular and unitarily similar to A, we have

$$\sum_{i=1}^{n}\sum_{j=1}^{n}|b_{ij}|^2 = \sum_{i=1}^{n}\left[\sum_{j=1}^{i-1}|b_{ij}|^2 + |b_{ii}|^2 + \sum_{j=i+1}^{n}|b_{ij}|^2\right] = \sum_{i=1}^{n}|\lambda_i|^2 + \sum_{i=1}^{n}\sum_{j=i+1}^{n}|b_{ij}|^2.$$

Hence

$$\sum_{i=1}^{n}\sum_{j=1}^{n}|a_{ij}|^2 \geq \sum_{i=1}^{n}|\lambda_i|^2.$$

This is *Schur's inequality*. Equality holds if and only if

$$\sum_{i=1}^{n}\sum_{j=i+1}^{n}|b_{ij}|^2 = 0,$$

i.e., if and only if B is a diagonal matrix.

Example 5.21. Schur's inequality can be applied to find a rough bound on the magnitudes of the individual eigenvalues. Since

$$|\lambda_i|^2 \leq \sum_{i=1}^{n}|\lambda_i|^2 \leq \sum_{i=1}^{n}\sum_{j=1}^{n}|a_{ij}|^2 \leq n^2 \max|a_{ij}|^2$$

we have $|\lambda_i| \leq n \max|a_{ij}|$ for $i = 1, \ldots, n$.   □

More interesting estimates for the eigenvalues are given by the next result.

Theorem 5.4 (Gershgorin's Theorem). *Each eigenvalue of a matrix A belongs to one of the circles centered at a_{ii} and of radius $\sum_{j \neq i}|a_{ij}|$, with $i = 1, \ldots, n$.*

Proof. Let λ be an eigenvalue to which there corresponds a column eigenvector $\mathbf{x}$. Let x_t be the component of $\mathbf{x}$ having maximal absolute value. Dividing $\mathbf{x}$ by x_t, we get an eigenvector that we again denote by $\mathbf{x}$. It has the form

$$\mathbf{x} = (x_1, \ldots, x_{t-1}, 1, x_{t+1}, \ldots, x_n)^T$$

and is such that $|x_k| \leq 1$ for all k. Now from the equality $A\mathbf{x} = \lambda\mathbf{x}$ we write down the equality for the tth component:

$$\sum_{j=1}^{n} a_{tj}x_j = \lambda x_t.$$

As $x_t = 1$, this can be rewritten as

$$\lambda - a_{tt} = \sum_{j \neq t} a_{tj}x_j$$

and by using the fact that $|x_k| \leq 1$, we get

$$|\lambda - a_{tt}| \leq \sum_{j \neq t} |a_{tj}|.$$

There are n such circles to one of which the eigenvalue λ should belong.   □

The theorem states that the circles are in $\mathbb{C}$, as A is not assumed Hermitian and so the eigenvalues can be complex.

Example 5.22. Schur's inequality can be used in conjunction with the AM–GM inequality to obtain a bound on the determinant of A. Recall that the determinant of a square matrix equals the product of its eigenvalues:

$$|\det A| = \left| \prod_{i=1}^{n} \lambda_i \right| = \prod_{i=1}^{n} |\lambda_i|.$$

Then

$$|\det A|^{2/n} = \left(\prod_{i=1}^{n} |\lambda_i|^2 \right)^{1/n} \leq \frac{1}{n} \sum_{i=1}^{n} |\lambda_i|^2 \leq \frac{1}{n} \sum_{i=1}^{n} \sum_{j=1}^{n} |a_{ij}|^2 \leq \frac{1}{n} n^2 \max |a_{ij}|^2,$$

and therefore

$$|\det A| \leq n^{n/2} (\max |a_{ij}|)^n.$$

This is the desired bound.   □

Inequalities also arise in the discussion of vector and matrix norms. We recall that for a column vector $\mathbf{x} = (x_1, \ldots, x_n)^T$, scalars called ℓ_p norms can be defined through the equation

$$\|\mathbf{x}\|_p = \left(\sum_{i=1}^{n} |x_i|^p \right)^{1/p} \qquad (p = 1, 2, \ldots, \infty).$$

In particular, $\|\mathbf{x}\|_2 = (\sum_{i=1}^{n} |x_i|^2)^{1/2}$ is the Euclidean or ℓ_2 norm. This norm has many applications in systems engineering, as does the ℓ_1 norm $\|\mathbf{x}\|_1 = \sum_{i=1}^{n} |x_i|$. The definition for $p = \infty$ is interpreted as $\|\mathbf{x}\|_\infty = \max |x_i|$.

Inequalities are available to interrelate the various vector norms. For instance, the reader might wish to verify that

$$\|\mathbf{x}\|_2 \leq \sqrt{n} \|\mathbf{x}\|_\infty, \qquad (\|\mathbf{x}\|_2)^2 \leq \|\mathbf{x}\|_1 \|\mathbf{x}\|_\infty.$$

For the vector norm $\|\mathbf{x}\|_p$, we define the induced matrix norm of the $n \times n$ matrix A by

$$\|A\|_p = \max_{\mathbf{x} \neq 0} \left\{ \frac{\|A\mathbf{x}\|_p}{\|\mathbf{x}\|_p} \right\}. \tag{5.31}$$

Then $\|A\mathbf{x}\|_p \leq \|A\|_p \|\mathbf{x}\|_p$ for all $\mathbf{x}$, and $\|A\mathbf{x}\|_p = \|A\|_p \|\mathbf{x}\|_p$ for some $\mathbf{x}$. Although $\|\mathbf{x}\|_2$ is the most natural way to measure the length of a vector, $\|A\|_2$ is difficult to compute in general. For this and other reasons, $p = 1$ or $p = \infty$ are frequent choices. $\|A\|_1$ gives the *max column sum*, i.e.,

$$\|A\|_1 = \max_{1 \le j \le n} \left\{ \sum_{i=1}^{n} |a_{ij}| \right\}$$

while $\|A\|_\infty$ gives the *max row sum*. Two other matrix norms that are commonly used are the Frobenius norm

$$\|A\|_F = \left(\sum_{i=1}^{n} \sum_{j=1}^{n} |a_{ij}|^2 \right)^{1/2} = (\mathrm{tr}[A^\dagger A])^{1/2}$$

and the cubic norm

$$\|A\|_C = n \max |a_{ij}|.$$

A property that must be satisfied by any valid matrix norm is

$$\|AB\| \le \|A\| \, \|B\| \tag{5.32}$$

for any two matrices A, B. It is easily verified that the Frobenius norm satisfies (5.32); by the Cauchy–Schwarz inequality,

$$\|AB\|_F = \left(\sum_{i=1}^{n} \sum_{j=1}^{n} \left| \sum_{k=1}^{n} a_{ik} b_{kj} \right|^2 \right)^{1/2} \le \left(\sum_{i=1}^{n} \sum_{j=1}^{n} \sum_{k=1}^{n} |a_{ik}|^2 \sum_{m=1}^{n} |b_{mj}|^2 \right)^{1/2}$$

$$= \left(\sum_{i=1}^{n} \sum_{k=1}^{n} |a_{ik}|^2 \sum_{j=1}^{n} \sum_{m=1}^{n} |b_{mj}|^2 \right)^{1/2} = \|A\|_F \, \|B\|_F \, .$$

The same property is also easily verified for the cubic norm. Another property of interest is that

$$\|A\|_2 \le \|A\|_F \le \sqrt{n} \, \|A\|_2 \, ,$$

providing an estimate for the more difficult to compute $\|A\|_2$.

A matrix norm is *compatible* with a given vector norm if the inequality

$$\|A\mathbf{x}\| \le \|A\| \, \|\mathbf{x}\|$$

holds for all $\mathbf{x}$. For example, $\|A\|_2$ and $\|A\|_F$ are both compatible with $\|\mathbf{x}\|_2$. The inequality from (5.31)

$$\|A\mathbf{x}\|_p \le \|A\|_p \, \|\mathbf{x}\|_p$$

is *sharp* in the sense that equality holds for some $\mathbf{x} \ne 0$. However, the Frobenius norm is not sharp (Problem 5.16). We pay a penalty for having an easily computed matrix norm $\|A\|_F$; it overestimates the "size" of the matrix A as an operator. Using any compatible norms, we can write

$$\|\lambda_i \mathbf{x}\| = |\lambda_i| \, \|\mathbf{x}\| = \|A\mathbf{x}\| \le \|A\| \, \|\mathbf{x}\|$$

and because the eigenvectors are nonzero,

$$|\lambda_i| \leq \|A\| \qquad (i = 1, \ldots, n).$$

The *spectral radius* $\rho[A]$ of the matrix A is defined to be the magnitude of the largest eigenvalue of A. Obviously then,

$$\rho[A] \leq \|A\|.$$

We have already met a specific instance of this inequality in Example 5.21, where the cubic norm was effectively used. See Theorem 6.9.2 of Stoer and Bulirsch [84] for more discussion on the spectral radius. See Marcus and Minc [53] for more discussion on matrix inequalities in general. Another good reference on matrices is Lütkepohl [51].

5.12 Topics in Signal Analysis

Consider a periodic square wave $w(t)$, given over one cycle by

$$w(t) = \begin{cases} -\pi/2 & \text{for } -\pi \leq t < 0, \\ \pi/2 & \text{for } \quad 0 \leq t < \pi. \end{cases}$$

By expressions given earlier, $w(t)$ can be represented as the Fourier series

$$w(t) = 2 \sum_{n=1}^{\infty} \frac{\sin(2n-1)t}{2n-1}. \tag{5.33}$$

The waveform $w(t)$ has a jump discontinuity at $t = 0$, and it is well known that a truncated version of the Fourier series will overshoot this jump (Fig. 5.4, the *Gibbs phenomenon*).

We now compute the amount of overshoot present (see [52]). Let us call the mth partial sum of the series $S_{wm}(t)$. Differentiation gives

$$\frac{dS_{wm}(t)}{dt} = 2 \sum_{n=1}^{m} \cos(2n-1)t.$$

Using the identity

$$\sum_{n=1}^{m} \cos(2n-1)t \equiv \frac{1}{2} \sin 2mt \csc t$$

and then integrating, we get

$$S_{wm}(t) = \int_0^t \frac{\sin 2m\tau}{\sin \tau} d\tau.$$

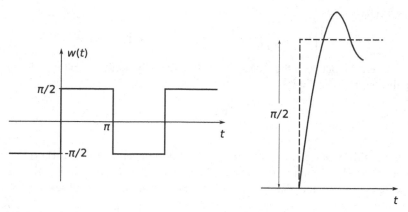

Fig. 5.4 Gibbs phenomenon in Fourier series. *Left*: Square wave $w(t)$. *Right*: Gibbs overshoot in a partial sum of the Fourier series. The *dashed line* indicates the initial portion of a positive-going cycle of $w(t)$

Defining

$$\Delta_m(t) = \left| S_{wm}(t) - \int_0^t \frac{\sin 2m\tau}{\tau} \, d\tau \right| \tag{5.34}$$

(the motivation becomes clear later) we see that

$$\Delta_m(t) = \left| \int_0^t \sin 2m\tau \, \frac{\tau}{\sin \tau} \left(\frac{1}{\tau} - \frac{\sin \tau}{\tau^2} \right) d\tau \right| \leq \int_0^t |\sin 2m\tau| \left| \frac{\tau}{\sin \tau} \right| \left| \frac{1}{\tau} - \frac{\sin \tau}{\tau^2} \right| d\tau.$$

But $\sin \tau \geq 2\tau/\pi$ for $0 \leq \tau \leq \pi/2$ by Jordan's inequality; also,

$$\frac{1}{\tau} - \frac{\sin \tau}{\tau^2} = \frac{\tau}{3!} - \frac{\tau^3}{5!} + \frac{\tau^5}{7!} - \cdots$$

so that for small positive τ we have

$$0 < \frac{1}{\tau} - \frac{\sin \tau}{\tau^2} < \frac{\tau}{3!}$$

and therefore

$$\Delta_m(t) \leq \int_0^t \frac{\pi}{2} \cdot \frac{\tau}{3!} \, d\tau = \frac{\pi}{24} t^2.$$

Hence for any $\varepsilon > 0$, there is a $T > 0$ such that $\Delta_m(t) < \varepsilon$ whenever $0 \leq t \leq T$. For $m > \pi/2T$, we may choose in particular $t = \pi/2m$ and after a change of variables write (5.34) as

$$\left| S_{wm}\left(\frac{\pi}{2m} \right) - \int_0^\pi \frac{\sin \tau}{\tau} \, d\tau \right| < \varepsilon. \tag{5.35}$$

But

$$\int_0^\pi \frac{\sin \tau}{\tau} \, d\tau = \int_0^\infty \frac{\sin \tau}{\tau} \, d\tau - \int_\pi^\infty \frac{\sin \tau}{\tau} \, d\tau = \frac{\pi}{2} - \int_\pi^\infty \frac{\sin \tau}{\tau} \, d\tau$$

and (5.35) is

$$\left| \left[S_{wm}\left(\frac{\pi}{2m}\right) - \frac{\pi}{2} \right] - \left[-\int_{\pi}^{\infty} \frac{\sin \tau}{\tau}\, d\tau \right] \right| < \varepsilon.$$

So the difference between the series and $w(t)$ in the neighborhood to the right of the jump discontinuity at $t = 0$ is

$$-\int_{\pi}^{\infty} \frac{\sin \tau}{\tau}\, d\tau \approx 0.281.$$

This is roughly 9 % of the jump height π and is independent of m. The fact that the Gibbs overshoot cannot be eliminated by a sufficiently large choice of m is, of course, related to the fact that the convergence of the series of functions in (5.33) is not uniform.

For aperiodic signals $f(t)$, the Fourier transform

$$\mathcal{F}[f(t)] = F(\omega) = \int_{-\infty}^{\infty} f(t)e^{-i\omega t}\, dt$$

and its inverse

$$\mathcal{F}^{-1}[F(\omega)] = f(t) = \frac{1}{2\pi} \int_{-\infty}^{\infty} F(\omega)e^{i\omega t}\, d\omega$$

are used to study frequency content as a function of the continuous angular frequency variable ω. By the differentiation property

$$\mathcal{F}\left[\frac{d^n f(t)}{dt^n}\right] = (i\omega)^n F(\omega) \tag{5.36}$$

we have

$$|\omega^n F(\omega)| = \left| \frac{1}{i^n} \int_{-\infty}^{\infty} \frac{d^n f}{dt^n} e^{-i\omega t}\, dt \right| \le \int_{-\infty}^{\infty} \left| \frac{d^n f}{dt^n} \right|\, dt$$

with the resulting bounds

$$|F(\omega)| \le \frac{1}{|\omega^n|} \int_{-\infty}^{\infty} \left| \frac{d^n f}{dt^n} \right|\, dt \qquad (n = 0, 1, 2, \ldots)$$

on the spectrum of f. This inequality supports our intuitive notion that only rapidly varying signals (i.e., signals having significant nth derivatives for large n) can have significant spectral content at high frequencies. A related fact is that short-duration time signals have broadband frequency spectra. In order to quantify this relationship, we use second-moment integrals to define the temporal duration of $f(t)$ as

$$D^2 = \int_{-\infty}^{\infty} t^2 f^2(t)\, dt$$

and the bandwidth of its spectrum as

$$B^2 = \int_{-\infty}^{\infty} \omega^2 |F(\omega)|^2\, d\omega.$$

The *uncertainty principle* states that if $f(t) = o(t^{-1/2})$, then

$$DB \geq \sqrt{\pi/2}. \tag{5.37}$$

To obtain (5.37) we write the Cauchy–Schwarz inequality for integrals as

$$\left| \int_{-\infty}^{\infty} [tf(t)] \left(\frac{df}{dt} \right) dt \right|^2 \leq \int_{-\infty}^{\infty} [tf(t)]^2 \, dt \int_{-\infty}^{\infty} \left(\frac{df}{dt} \right)^2 dt. \tag{5.38}$$

But integration by parts gives

$$\int_{-\infty}^{\infty} [tf(t)] \left(\frac{df}{dt} \right) dt = \int_{-\infty}^{\infty} tf(t) \, df(t) = t \frac{f^2(t)}{2} \bigg|_{-\infty}^{\infty} - \int_{-\infty}^{\infty} \frac{f^2(t)}{2} \, dt,$$

where the first term in the rightmost member vanishes by the o condition on f. The integral

$$E = \int_{-\infty}^{\infty} f^2(t) \, dt$$

is called the normalized energy in the signal f; without loss of generality this can be set to unity to give

$$\frac{1}{4} \leq D^2 \int_{-\infty}^{\infty} \left(\frac{df}{dt} \right)^2 dt.$$

It only remains to invoke (5.36) and Parseval's identity in the form

$$2\pi \int_{-\infty}^{\infty} f^2(t) \, dt = \int_{-\infty}^{\infty} |F(\omega)|^2 \, d\omega$$

to obtain (5.37). It is easily shown (Problem 5.17) that the minimum duration-bandwidth product is realized [i.e., equality is attained in (5.37)] when the signal f is a Gaussian pulse.

5.13 Dynamical System Stability and Control

A broad class of continuous-time systems can be modeled using an initial-value problem of the form

$$\begin{cases} \dfrac{d\mathbf{x}(t)}{dt} = f(\mathbf{x}(t)) & (t \geq t_0), \\ \mathbf{x}(t_0) = \mathbf{x}_0, \end{cases}$$

where $\mathbf{x}(t)$ is the N-dimensional state vector of the system, and the system structure is reflected in the function f. Such systems are unforced (i.e., no input signal). The set of all possible $\mathbf{x}$ is the state space of the system, and solution curves in state space are known as system trajectories.

Stability theory deals with sensitivity to unwanted disturbances. Of special concern are disturbances tending to perturb the system from an *equilibrium state*, a value $\mathbf{x} = \mathbf{x}_e$ such that $f(\mathbf{x}_e) = \mathbf{0}$ whenever $t \geq t_0$. Because such a state may always be transferred to the origin of state space by a suitable coordinate translation, it is customary to take $\mathbf{x}_e = \mathbf{0}$. If $\mathbf{x}_e$ is unstable, a slight perturbation could put the system on a trajectory leading away from $\mathbf{x}_e$; on the other hand, the trajectory could stay within a small neighborhood of $\mathbf{x}_e$, or it could lead back to $\mathbf{x}_e$. Technically, there are several notions of stability. The origin $\mathbf{x}_e = \mathbf{0}$ is ...

(a) *stable in the sense of Lyapunov* if for every $\varepsilon > 0$ there exists $\delta(\varepsilon) > 0$ such that if $\|\mathbf{x}(t_0)\| < \delta$, then the resulting trajectory satisfies $\|\mathbf{x}(t)\| < \varepsilon$ for all $t > t_0$.

(b) *asymptotically stable* if it is stable in the sense of Lyapunov and, in addition, there exists $\gamma > 0$ such that whenever $\|\mathbf{x}(t_0)\| < \gamma$ the resulting trajectory has $\|\mathbf{x}(t)\| \to 0$ as $t \to \infty$.

(c) *exponentially stable* if there exist positive numbers α, λ, such that for all $t > t_0$, $\|\mathbf{x}(t)\| \leq \alpha \|\mathbf{x}(t_0)\| e^{-\lambda t}$ whenever $\mathbf{x}(t_0)$ lies sufficiently close to $\mathbf{x}_e$.

Lyapunov theory can yield conclusions about stability without explicit knowledge of $\mathbf{x}(t)$. This theory is extensive, and here we can offer only a few preliminary remarks for the reader. A principal notion is that if a system having just one equilibrium state is dissipative, then the system will always return to that state after any perturbation from it. This equilibrium will be a point of minimum energy for the system, and as any trajectory is followed toward this point the system energy must continually decrease. Use is therefore made of a "generalized energy" function $V(\mathbf{x})$, called a *Lyapunov function*. Assume $V(\mathbf{x})$ is continuous with continuous partial derivatives in state space, and let Ω be a region about $\mathbf{x} = \mathbf{0}$. We say that $V(\mathbf{x})$ is *positive definite* in Ω if

(1) $V(\mathbf{0}) = 0$, and
(2) $V(\mathbf{x}) > 0$ for every nonzero $\mathbf{x} \in \Omega$.

$V(\mathbf{x})$ is *negative semidefinite* in Ω if $V(\mathbf{0}) = 0$ and $V(\mathbf{x}) \leq 0$ for every nonzero $\mathbf{x} \in \Omega$. Similar definitions can be formulated for the terms positive semidefinite and negative definite. We can now state a simple stability result.

Theorem 5.5. *If a positive definite function $V(\mathbf{x})$ can be determined for a system such that dV/dt is negative semidefinite, then the equilibrium point $\mathbf{x} = \mathbf{0}$ is stable in the sense of Lyapunov. Here dV/dt means $dV(\mathbf{x}(t))/dt$, which is also written as $dV(\mathbf{x})/dt$.*

Proof. Let $\varepsilon > 0$ be given and write $S_\varepsilon = \{\mathbf{x} : \|\mathbf{x}\| = \varepsilon\}$. Because S_ε is closed and bounded, as in the case for one variable, $V(\mathbf{x})$ assumes a minimum value m on S_ε. Note that $m > 0$ because $V(\mathbf{x})$ is positive definite about $\mathbf{x} = \mathbf{0}$. Continuity of V at $\mathbf{x} = \mathbf{0}$ guarantees a $\delta > 0$ such that $\delta < \varepsilon$ and $V(\mathbf{x}) \leq m/2$ whenever $\|\mathbf{x}\| \leq \delta$. Also, because $dV(\mathbf{x})/dt$ is negative semidefinite we know that $V(\mathbf{x}(t))$ is nonincreasing with respect to t. Hence, for an initial condition with $\|\mathbf{x}(t_0)\| \leq \delta$, we have $V(\mathbf{x}(t)) \leq V(\mathbf{x}(t_0)) \leq m/2$ for $t > t_0$. But this implies that $\|\mathbf{x}(t)\| < \varepsilon$ for

$t > t_0$. Indeed, suppose to the contrary that $\|\mathbf{x}(t)\| \geq \varepsilon$ at some time $t > t_0$. Because $\|\mathbf{x}(t_0)\| < \delta < \varepsilon$, at some intermediate time $t_0 < t' \leq t$ we have $\|\mathbf{x}(t')\| = \varepsilon$. But on S_ε we have $V(\mathbf{x}) \geq m$, contradicting $V(\mathbf{x}(t)) \leq m/2$ for $t > t_0$.   □

Moreover, if $dV(\mathbf{x})/dt$ is negative definite, then $\mathbf{x} = \mathbf{0}$ is asymptotically stable. Geometrically, we may imagine a level contour or surface $V(\mathbf{x}) = $ constant > 0 (Fig. 5.5). Let $\mathbf{x}$ be on this contour or surface. Since $\mathbf{x} \neq \mathbf{x}_e$ we require $dV(\mathbf{x})/dt < 0$. By the chain rule,

$$\frac{dV(\mathbf{x})}{dt} = (\nabla V(\mathbf{x}))^T f(\mathbf{x}), \tag{5.39}$$

$\nabla V(\mathbf{x})$ represents an outward normal to the level contour or surface, and the vector field $f(\mathbf{x})$ provides tangent vectors along solutions. In engineering terminology, $(\nabla V(\mathbf{x}))^T f(\mathbf{x})$ is the dot product, the product of the magnitudes (norms) with the cosine of the angle between the two vectors. The fact that the cosine of the angle is negative means that the solution is directed inward, hence a solution starting in the interior of the level contour or surface can never leave the interior and so remains bounded for all time and, in fact, approaches $\mathbf{x}_e$ as $t \to \infty$. The Lyapunov function $V(\mathbf{x})$ sometimes corresponds to actual physical energy, but not always.

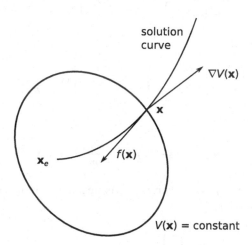

Fig. 5.5 Lyapunov function

Example 5.23. Consider the motion of a mass m attached to a nonlinear spring with stiffness $k(x + x^3)$, where x is the distance from the equilibrium position, and a dashpot (shock absorber) is attached to provide a damping force $c(dx/dt)$. The differential equation is

$$mx'' + cx' + k(x + x^3) = 0.$$

The equation is converted to a first-order system by substituting $x' = y$, so our system is

$$\frac{dx}{dt} = y, \qquad \frac{dy}{dt} = -\frac{k}{m}(x + x^3) - \frac{c}{m}y.$$

The kinetic and potential energies are given respectively by

$$K = \frac{1}{2}mv^2 = \frac{1}{2}my^2, \qquad P = \int k(x + x^3)\,dx = k\left(\frac{x^2}{2} + \frac{x^4}{4}\right).$$

Hence the total energy should be a candidate for the Lyapunov function

$$V\begin{pmatrix} x \\ y \end{pmatrix} = k\left(\frac{x^2}{2} + \frac{x^4}{4}\right) + \frac{1}{2}my^2. \tag{5.40}$$

By (5.39),

$$\frac{dV}{dt}\begin{pmatrix} x \\ y \end{pmatrix} = \left(\nabla V\begin{pmatrix} x \\ y \end{pmatrix}\right)^T f\begin{pmatrix} x \\ y \end{pmatrix} = -cy^2. \tag{5.41}$$

Equations (5.40) and (5.41) show that V is positive definite and dV/dt is negative semidefinite, so the system is stable. However, physical intuition tells us the damped system should be asymptotically stable. Unfortunately, for this choice of V the derivative is zero along the x-axis, and we wanted

$$\frac{dV}{dt}\begin{pmatrix} x \\ y \end{pmatrix} < 0 \qquad \text{if} \qquad \begin{pmatrix} x \\ y \end{pmatrix} \neq \begin{pmatrix} 0 \\ 0 \end{pmatrix}.$$

After some fiddling, let

$$V\begin{pmatrix} x \\ y \end{pmatrix} = k\left(\frac{x^2}{2} + \frac{x^4}{4}\right) + \frac{1}{2}my^2 + \beta\left(xy + \frac{c}{m}\frac{x^2}{2}\right).$$

Then, using (5.39) again,

$$\frac{dV}{dt}\begin{pmatrix} x \\ y \end{pmatrix} = (-c + \beta)y^2 - \frac{\beta k}{m}(x^2 + x^4).$$

Thus, if we choose $0 < \beta < c$, then dV/dt is negative definite. We want V to be positive definite. Rewrite

$$V\begin{pmatrix} x \\ y \end{pmatrix} = \frac{kx^4}{4} + W\begin{pmatrix} x \\ y \end{pmatrix}.$$

Recognize

$$W\begin{pmatrix} x \\ y \end{pmatrix} = \left(\frac{k}{2} + \frac{\beta c}{2m}\right)x^2 + \beta xy + \frac{m}{2}y^2$$

as the quadratic form

$$W\begin{pmatrix} x \\ y \end{pmatrix} = \begin{pmatrix} x \\ y \end{pmatrix}^T \begin{pmatrix} a_{11} & a_{12} \\ a_{21} & a_{22} \end{pmatrix}\begin{pmatrix} x \\ y \end{pmatrix},$$

where $a_{11} = k/2 + \beta c/2m$, $a_{12} = a_{21} = \beta/2$, and $a_{22} = m/2$. Recall from Theorem 5.2 that if A is symmetric and

$$a_{11} > 0 \text{ and the determinant } \begin{vmatrix} a_{11} & a_{12} \\ a_{21} & a_{22} \end{vmatrix} > 0, \tag{5.42}$$

then W is positive definite. It is easily verified that the choice $\beta = c/2$ guarantees that (5.42) is satisfied. Hence W and therefore V is positive definite. Because $0 < \beta < c$ we have guaranteed that dV/dt is negative definite. In other words, the system is asymptotically stable at the origin. In this relatively simple example finding a Lyapunov function was not immediate. Before continuing we simplify matters by assuming $m = k = c = 1$ and $\beta = 1/2$. By Theorem 5.1 the quadratic form W satisfies

$$\lambda_1(x^2 + y^2) \leq W \leq \lambda_2(x^2 + y^2),$$

where the eigenvalues of A are

$$\lambda_1 = \frac{5 - \sqrt{5}}{8} \quad \text{and} \quad \lambda_2 = \frac{5 + \sqrt{5}}{8}.$$

Since $V = (x^4/4) + W$,

$$\frac{x^4}{4} + \lambda_1(x^2 + y^2) \leq V \leq \frac{x^4}{4} + \lambda_2(x^2 + y^2). \tag{5.43}$$

Now

$$\frac{dV}{dt} = -\frac{x^2 + y^2}{2} - \frac{x^4}{2}$$

so we may substitute for $x^2 + y^2$ to get

$$V \leq \frac{x^4}{4} + \lambda_2 \left(-2\frac{dV}{dt} - x^4 \right) \leq -2\lambda_2 \frac{dV}{dt},$$

which implies

$$V(t) \leq V(0) e^{-t/2\lambda_2}.$$

We use (5.43) and observe

$$\lambda_1 x^2 \leq \frac{x^4}{4} + \lambda_1(x^2 + y^2) \leq V,$$

hence

$$|x(t)| \leq \sqrt{\frac{V(0)}{\lambda_1}} e^{-t/4\lambda_2}.$$

Similarly, $|y(t)|$ and therefore $\|(x, y)^T\|$ are bounded by decaying exponentials and the system is exponentially stable at the origin. The reader interested in pursuing Lyapunov theory further is invited to consult [11, 40, 41]. □

When considering systems with nonzero inputs, other notions of stability must be employed. A crucial question is whether a bounded input signal will always give rise to a bounded output signal. If so, the system is said to have *bounded-input, bounded-output* (BIBO) *stability*. The system depicted in Fig. 5.6 is BIBO stable provided there is a constant I such that if $|u(t)| \le B$ for all t, then $|y(t)| \le BI$ for all t.

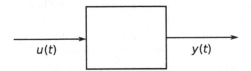

Fig. 5.6 Block diagram of single-input, single-output system

A linear, time-invariant (LTI) system can be modeled by an equation of the form

$$L[y] = u$$

where the operator L is time-independent and linear. For instance, L could be a linear constant-coefficient differential operator

$$L = a_n \frac{d^n}{dt^n} + \cdots + a_1 \frac{d}{dt} + a_0.$$

For a relaxed LTI system (i.e., a system having zero initial conditions), there is a function $h(t)$ such that

$$y(t) = h(t) * u(t) = \int_0^\infty h(\tau) u(t - \tau) \, d\tau.$$

Hence, knowledge of $h(t)$ is sufficient to determine the output $y(t)$ produced by a given input $u(t)$. We take the system to be causal so that $h(t) \equiv 0$ whenever $t < 0$. For bounded inputs

$$|y(t)| \le \int_0^\infty |h(\tau)| \, |u(t - \tau)| \, d\tau \le B_1 \int_0^\infty |h(\tau)| \, d\tau,$$

and thus a sufficient condition for BIBO stability is that $h(t)$ be absolutely integrable on $[0, \infty)$:

$$\int_0^\infty |h(t)| \, dt = B_2 < \infty.$$

Conversely, suppose the system has BIBO stability. In particular, if we choose $B = 1$ there exists a constant M such that if the input is bounded by B, the output is bounded by M. We claim that

$$\int_0^\infty |h(t)| \, dt < \infty.$$

If not, choose T so that

$$\int_0^T |h(t)| \, dt > M.$$

Define the bounded input

$$u(t) = \begin{cases} \dfrac{h(T - t)}{|h(T - t)|}, & 0 \le t \le T, \quad h(T - t) \ne 0, \\ \\ 0, & \text{otherwise.} \end{cases}$$

Then

$$y(T) = \int_0^T h(\tau)u(T - \tau)\, d\tau = \int_0^T |h(\tau)|\, d\tau > M,$$

which is a contradiction.

An alternative system description often employed for relaxed LTI systems is the transfer function $H(s)$, defined as the Laplace transform of $h(t)$:

$$H(s) = \int_0^\infty h(t)e^{-st}\, dt.$$

Again, this function should contain all necessary information about the system. As a function of the complex variable s, however, its interesting properties involve its singularities in the complex plane. Of these, the poles (the values of s for which $|H(s)| \to \infty$) are of greatest importance. To see why, let s_0 be a point located either on the imaginary axis or in the right-half of the s-plane so that $\text{Re}[s_0] \ge 0$. Then for any $t \ge 0$ we have $|e^{-s_0 t}| \le 1$ and

$$|H(s_0)| \le \int_0^\infty |h(t)|\, dt = B_2,$$

provided the system is BIBO stable. Hence stability implies that all the poles of $H(s)$ must lie in the left-half of the s-plane, a result familiar to electrical and mechanical engineers.

Like stability, the subject of control is huge. We give one example.

Example 5.24. The differential equation

$$\frac{d\omega(t)}{dt} + a\omega(t) = Kv(t)$$

can model the angular shaft speed $\omega(t)$ of a fixed-field, armature-controlled dc motor. The forcing function $v(t)$ is the voltage applied to the armature, while a and K are positive constants describing the resistance of the motor windings, rotational inertial of the shaft and its load, frictional effects, back emf, etc. Application of Laplace transform methods, with zero initial conditions on the shaft speed, yields

$$h(t) = Ke^{-at} \quad \text{for } t > 0,$$

and hence by convolution with the input

$$\omega(t) = \int_0^t v(\lambda)h(t - \lambda)\, d\lambda = \int_0^t v(\lambda)Ke^{-a(t-\lambda)}\, d\lambda$$

for all $t > 0$. A simple motor control question is this: What input $v(t)$ should be applied in order to bring the shaft from rest to some given speed ω_d, in time T, while keeping the input energy integral

$$E_v = \int_0^T |v(t)|^2 \, dt$$

a minimum? By the Cauchy–Schwarz inequality,

$$\omega_d^2 \le E_v \int_0^T K^2 e^{-2a(T-\lambda)} \, d\lambda$$

and hence the energy required for the task satisfies

$$E_v \ge \frac{2a\omega_d^2}{K^2(1 - e^{-2aT})}.$$

Equality is attained when

$$v(\lambda) = K_p e^{-a(T-\lambda)},$$

where the proportionality constant K_p is determined by setting

$$\omega_d = \int_0^T K_p e^{-a(T-\lambda)} K e^{-a(T-\lambda)} \, d\lambda = \frac{K_p K}{2a} (1 - e^{-2aT}).$$

Hence the optimal driving voltage waveform is given for $t > 0$ by

$$v(t) = \frac{2a\omega_d}{K(1 - e^{-2aT})} e^{-a(T-t)}.$$

Putting this back into the convolution integral, we write the resulting shaft speed as

$$\omega(t) = \omega_d \frac{e^{at} - e^{-at}}{e^{aT} - e^{-aT}} = \omega_d \frac{\sinh at}{\sinh aT}.$$

$\square$

5.14 Some Inequalities of Probability

Take a random variable $X \ge 0$. Then for any $t > 0$, the probability of the event $X \ge t$ satisfies

$$P(X \ge t) \le \frac{\mu_X}{t}, \tag{5.44}$$

where μ_X is the mean or expected value of X. This is the *Markov inequality*. To illustrate how it is developed, let us consider the case where X is a discrete random variable having frequency function $f_X(x) = P(X = x) \geq 0$ (recall that probabilities are always nonnegative). We have

$$\mu_X = \sum_{x \geq 0} x f_X(x) = \sum_{0 \leq x < t} x f_X(x) + \sum_{x \geq t} x f_X(x),$$

so that

$$\mu_X \geq \sum_{x \geq t} x f_X(x) \geq t \sum_{x \geq t} f_X(x) = t P(X \geq t),$$

and (5.44) follows. The development for a continuous random variable is similar.

The inequality

$$P(|X - \mu_X| \geq t) \leq \frac{\sigma_X^2}{t^2}, \tag{5.45}$$

where σ_X^2 is the variance of X, is the *Chebyshev inequality*. To derive it from the Markov inequality, we start from the obvious fact that $(X - \mu_X)^2 \geq 0$. Then, for every $t > 0$,

$$P\left((X - \mu_X)^2 \geq t^2\right) \leq \frac{E[(X - \mu_X)^2]}{t^2} = \frac{\sigma_X^2}{t^2}.$$

But $(X - \mu_X)^2 \geq t^2$ if and only if $|X - \mu_X| \geq t$, and (5.45) follows.

Example 5.25. The special case $P(|X - \mu_X| \geq n\sigma_X) \leq 1/n^2$ shows that a random variable X is likely to fall close to its mean. ☐

Example 5.26. Binomially distributed random variables have mean np and variance $np(1 - p)$, where n is the number of trials of the experiment and p is the probability of "success" in each trial. Then $t = n\beta$ gives

$$P(|X - np| < n\beta) \geq 1 - \frac{p(1 - p)}{n\beta^2}.$$

For instance, suppose a population contains an unknown proportion p of defective objects. Let X be the number of defectives in a sample of size N. Then for every $\beta > 0$,

$$P\left(\left|\frac{X}{N} - p\right| < \beta\right) \geq 1 - \frac{p(1 - p)}{N\beta^2}.$$

Now $\max[p(1 - p)] = 0.25$; hence for fixed β, N, the minimum probability that the observed proportion of defectives in the sample differs from the actual proportion p by an amount less than β is $1 - 0.25/N\beta^2$. Hence, to insure that this probability meets or exceeds some value P, we need $N \geq 0.25/(1 - P)\beta^2$. ☐

As the Markov inequality (5.44) requires knowledge of only the mean of a random variable, it tends to provide loose bounds. When $t < \mu_X$, it merely gives $P(X \geq t) \leq 1$. Furthermore, the inequality applies only to nonnegative random variables. However, if s is a real parameter then e^{sX} is nonnegative for any random variable s. This observation will enable us to apply (5.44) to an arbitrary random variable X and then adjust the parameter s to achieve some desired outcome.

Take $s > 0$. By monotonicity of the exponential function we have $X \geq t$ if and only if $e^{sX} \geq e^{st}$, and (5.44) gives

$$P(X \geq t) = P(e^{sX} \geq e^{st}) \leq \frac{E[e^{sX}]}{e^{st}}$$

or

$$P(X \geq t) \leq e^{-st} E[e^{sX}]. \tag{5.46}$$

Similarly, for $s < 0$ we obtain

$$P(X \leq t) \leq e^{-st} E[e^{sX}]. \tag{5.47}$$

Inequalities (5.46) and (5.47) provide a family of bounds, indexed by parameter s, for the tail probabilities of X. These are known as *Chernoff bounds*. We may select s to obtain, for instance, a convenient expression for the bound. Alternatively, we may seek the tightest available bound; in the case of (5.46) we have

$$P(X \geq t) \leq \inf_{s>0} e^{-st} E[e^{sX}]. \tag{5.48}$$

We may be able to minimize $e^{-st} E[e^{sX}]$ by differentiation. The quantity $E[e^{sX}]$ is known as the moment generating function of X.

Example 5.27. Let X have the standard normal distribution. For this random variable it can be shown that $E[e^{sX}] = e^{s^2/2}$. Setting

$$\frac{d}{ds} \left[e^{-st+s^2/2} \right] = 0$$

we get $s = t$ and hence $P(X \geq t) \leq e^{-t^2/2}$.   □

Many important continuous random variables are normally distributed. Recall that the standard normal density has

$$f_X(x) = \frac{1}{\sqrt{2\pi}} e^{-x^2/2}$$

for $-\infty < x < \infty$. It is often convenient to work with the related coerror function

$$Q(x) = P[X > x] = \frac{1}{\sqrt{2\pi}} \int_x^\infty e^{-t^2/2} \, dt$$

for which repeated integration by parts yields the asymptotic expansion [1]

$$Q(x) = \frac{e^{-x^2/2}}{\sqrt{2\pi}x} \left[1 - \frac{1}{x^2} + \frac{1 \cdot 3}{x^4} + \cdots + \frac{(-1)^n \cdot 1 \cdot 3 \cdot 5 \cdots (2n-1)}{x^{2n}} \right]$$

$$+ \frac{(-1)^{n+1} \cdot 1 \cdot 3 \cdot 5 \cdots (2n+1)}{\sqrt{2\pi}} \int_x^\infty \frac{e^{-t^2/2}}{t^{2n+2}} \, dt \qquad (x > 0).$$

The inequalities

$$\frac{1}{\sqrt{2\pi}x}\left(1 - \frac{1}{x^2}\right)e^{-x^2/2} < Q(x) < \frac{1}{\sqrt{2\pi}x}e^{-x^2/2} \qquad (5.49)$$

are thus apparent. Tighter bounds on $Q(x)$ are also available (see [9] and Problem 5.18). This function is of interest in communication engineering, where it is often assumed that the system noise is Gaussian. We touch on some other aspects of communication systems in the next section.

5.15 Applications in Communication Systems

The field of communications strives toward the accurate reception of information signals in the presence of noise. Noise phenomena can be typed according to their physical mode of production, or simply according to the shape of their power spectra. For instance, much noise is produced by electrons moving randomly in conductors; this noise, called thermal noise, is approximately *white* because power in the associated waveform tends to be distributed evenly across the frequency spectrum.

In binary communication, time is divided into successive bit intervals, of length T seconds say, and during each interval at the receiver a known deterministic signal $g_i(t)$ is either present or absent (corresponding to binary 1 or binary 0). Of course, noise $n_i(t)$ is present in either case (subscript i denotes waveforms at the receiver input, as in Fig. 5.7). Since the form of $g_i(t)$ is known in advance, the receiver's function is simply to make a presence/absence decision during each bit interval. The decision process is, of course, complicated by $n_i(t)$, which in many cases enters in additive fashion so that the received waveform is $g_i(t)+n_i(t)$. An error occurs if the receiver decides $g_i(t)$ is present when it was never transmitted, or vice versa. It can be shown that the probability of such an error is minimized if the decision is based on a sample taken from the received waveform at a time instant when the signal-to-noise power ratio S/N is maximum. Hence we become interested in a system block that can enhance signal power at some instant of time while simultaneously reducing average noise power. Such a device, known as a *matched filter*, can be found as follows.

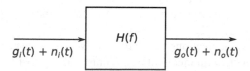

Fig. 5.7 Pre-decision filter

We assume additive white noise with power spectral density $N_0/2$ (watts per Hz), and seek an expression for the Fourier-domain transfer function $H(f)$. The signal output from the filter is given by Fourier inversion as

$$g_0(t) = \mathcal{F}^{-1}[G_0(f)] = \int_{-\infty}^{\infty} G_i(f)H(f)e^{i2\pi ft}\, df,$$

so that at the sample time $t = T$ its normalized power is

$$S = g_0^2(T) = \left| \int_{-\infty}^{\infty} G_i(f)H(f)e^{i2\pi fT}\, df \right|^2.$$

The power spectral density of the output noise is given by $(N_0/2)|H(f)|^2$ because the power gain of the linear system $H(f)$ at frequency f is $|H(f)|^2$; hence the normalized output average noise power is

$$N = \int_{-\infty}^{\infty} (N_0/2)|H(f)|^2\, df$$

and the signal-to-noise ratio is

$$\frac{S}{N} = \frac{\left| \int_{-\infty}^{\infty} G_i(f)H(f)e^{i2\pi fT}\, df \right|^2}{\int_{-\infty}^{\infty}(N_0/2)|H(f)|^2\, df}.$$

The integral in the numerator can be expressed in inner-product form (see Example 4.10):

$$\int_{-\infty}^{\infty} G_i(f)H(f)e^{i2\pi fT}\, df = \int_{-\infty}^{\infty} H(f)\overline{G_i(f)}e^{-i2\pi fT}\, df.$$

But by the Cauchy–Schwarz inequality (see Example 4.11),

$$\frac{S}{N} \le \frac{\int_{-\infty}^{\infty}|G_i(f)|^2\, df \int_{-\infty}^{\infty}|H(f)|^2\, df}{(N_0/2)\int_{-\infty}^{\infty}|H(f)|^2\, df}.$$

Equality is attained when $H(f)$ is proportional to $\overline{G_i}(f)e^{-i2\pi fT}$, leading to the choice

$$h(t) = \mathcal{F}^{-1}\{\overline{G_i}(f)e^{-i2\pi fT}\} = \overline{g}_i(T - t)$$

for the matched filter. For more details on the matched filter, along with derivation of the optimal error rate expressions in terms of coerror function $Q(x)$, the reader is referred to Couch [16].

The study of communications naturally involves some aspects of *information theory*, a subject replete with inequalities beginning at its most elementary level. To formally define such a nebulous concept as "information" must have been a substantial challenge to Claude Shannon and the other early workers in this area. To help make the idea precise, we imagine a hypothetical machine called a discrete memoryless source (DMS). The DMS has an alphabet ζ, which is just a discrete set of symbols $\zeta = \{S_1, \ldots, S_N\}$, and it periodically emits one of these symbols S as a *message* to the outside world (Fig. 5.8). The symbols are emitted randomly with

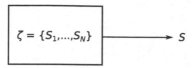

Fig. 5.8 An information source

probabilities $P(S = S_n) = p_n$ (such that $\sum p_n = 1$). These probabilities are assumed to be time-independent, and successive symbols are taken to be statistically independent. The *self-information* associated with each symbol is defined as

$$I_n = \log_b(1/p_n) \qquad (n = 1, \ldots, N).$$

This definition passes some key intuitive tests. Because $0 \le p_n \le 1$, we have $I_n \ge 0$, and need not worry about the possibility of getting "negative information" from the source. As $p_n \to 1$, $I_n \to 0$; that is, a symbol that occurs with such stubborn repetitiveness as to be deterministic would never surprise an observer and should carry no information. We have $I_n > I_m$ whenever $p_n < p_m$; the monotonic behavior of the logarithm means that unlikely symbols carry more information than likely ones. Finally, the joint probability $P(S_n \text{ and } S_m)$ equals $p_n p_m$ for two successive independent messages, leading to a joint information quantity satisfying

$$I_{nm} = \log_b[1/(p_n p_m)] = \log_b(1/p_n) + \log_b(1/p_m) = I_n + I_m.$$

The logarithmic base b is arbitrary but determines the unit of information. The usual choice $b = 2$ gives information in bits. Because the self-information I is a random variable taking possible values $I_1, \ldots, I_N$, we can compute its expected value as

$$H(\zeta) = \sum_{n=1}^{N} I_n p_n = \sum_{n=1}^{N} p_n \log_2(1/p_n).$$

This quantity, the average information per symbol, is the *entropy* of the DMS. Of special interest are bounds on $H(\zeta)$. Certainly

$$H(\zeta) \ge 0$$

and equality holds if and only if all the p_n except one vanish (again the case of no uncertainty, no information). For an upper bound we may convert to the natural log via the formula $\log_2 x = K \ln x$ (the constant K is immaterial) and use

$$\ln x \le x - 1$$

(Problem 2.7), which is regarded as a fundamental inequality of information theory. We have

$$\log_2 N = \log_2 N \sum_{n=1}^{N} p_n = \sum_{n=1}^{N} p_n \log_2 N$$

so that

$$H(\zeta) - \log_2 N = \sum_{n=1}^{N} p_n[\log_2(1/p_n) - \log_2 N] = \sum_{n=1}^{N} p_n \log_2[1/(Np_n)]$$

$$= K \sum_{n=1}^{N} p_n \ln[1/(Np_n)] \leq K \sum_{n=1}^{N} p_n[1/(Np_n) - 1].$$

The last quantity vanishes, hence

$$H(\zeta) \leq \log_2 N.$$

This upper bound is attained if and only if $1/(Np_n) = 1$ for all n, that is, when $p_n = 1/N$ for all n. The entropy of a DMS is therefore greatest when its output is least predictable on average, i.e., when all the message outputs are equally probable.

As usual, we could only touch on a few preliminary aspects of this fascinating subject here. The interested reader can consult [8, 44, 82] for more.

5.16 Existence of Solutions

Theorem 4.3, the contraction mapping theorem, is a powerful result that allows us to prove the existence of unique solutions to many equations of practical importance. In addition, the proofs yield practical methods such as Neumann series and Picard iteration for solving certain integral equations and differential equations, and Newton's method for solving systems of nonlinear equations.

Integral equations, where the unknown is a function appearing underneath an integral sign, arise naturally in areas like mechanics, electromagnetics, control, and population dynamics. For example, the equation

$$\psi(x) = g(x) + \lambda \int_a^b K(x,t)\psi(t)\,dt \qquad (a \leq x \leq b), \tag{5.50}$$

where $\psi(x)$ is unknown, is called a Fredholm integral equation of the second kind. We assume that $g(x) \in C[a,b]$, and that the kernel $K(x,t)$ is continuous for both $a \leq x \leq b$ and $a \leq t \leq b$. We seek a condition under which the integral operator

$$F(\psi)(x) = g(x) + \lambda \int_a^b K(x,t)\psi(t)\,dt$$

is a contraction on $C[a,b]$. Now since $K(x,t)$ is continuous on a closed and bounded domain, we know that $K(x,t)$ is bounded (by B, say). Let $u(x)$ and $v(x)$ be arbitrary members of $C[a,b]$. Then

$$d(F(u), F(v)) = \max_{x\in[a,b]} \left| \lambda \int_a^b K(x,t)[u(t) - v(t)] \, dt \right|$$

$$\leq \max_{x\in[a,b]} |\lambda| \int_a^b |K(x,t)| \, |u(t) - v(t)| \, dt$$

$$\leq B|\lambda| \max_{x\in[a,b]} \int_a^b |u(t) - v(t)| \, dt$$

$$\leq B|\lambda| \, (b-a) \max_{x\in[a,b]} |u(x) - v(x)|$$

$$= B|\lambda| \, (b-a) \, d(u(x), v(x)).$$

For F to be a contraction on $C[a,b]$ then, we require that

$$|\lambda| < 1/B(b-a).$$

Provided this condition is satisfied, we may iterate to solve (5.50) for $\psi(x)$ on $[a,b]$. Starting with an initial guess of $\psi^{(0)}(x) = g(x)$, the first iteration yields

$$\psi^{(1)}(x) = g(x) + \lambda \int_a^b K(x,t)\psi^{(0)}(t) \, dt = g + \lambda \Gamma[g],$$

where for convenience we have defined Γ as the integral operator

$$\Gamma[\psi] = \int_a^b K(x,t)\psi(t) \, dt.$$

The second iteration is then

$$\psi^{(2)} = g + \lambda \Gamma[g] + \lambda \Gamma^2[g] = g + \sum_{i=1}^{2} \lambda^i \Gamma^i[g]$$

and, in general,

$$\psi^{(n)} = g + \sum_{i=1}^{n} \lambda^i \Gamma^i[g].$$

By Theorem 4.3, we can express the solution of (5.50) as

$$\psi = \lim_{n\to\infty} \psi^{(n)} = g + \sum_{i=1}^{\infty} \lambda^i \Gamma^i[g].$$

This is the *Neumann series* for the integral equation.

Example 5.28. A specific instance of (5.50) is furnished by

$$\psi(x) = g(x) + \lambda \int_0^1 e^{x-t}\psi(t) \, dt \qquad (0 \leq x \leq 1). \qquad (5.51)$$

In this case it is easily verified that $\Gamma^n[g] = \kappa e^x$ for $n \in \mathbb{N}$, where

$$\kappa = \int_0^1 e^{-t}g(t) \, dt.$$

Hence

$$\psi(x) = g(x) + \sum_{i=1}^{\infty} \lambda^i \kappa e^x = g(x) + \kappa e^x \frac{\lambda}{1 - \lambda} \tag{5.52}$$

for $0 \le x \le 1$, provided that $|\lambda| < 1$. The validity of (5.52) as a solution to (5.51) is easily verified by direct substitution.   □

The reader interested in pursuing integral equations further could refer to Jerri [39]. Another application of Theorem 4.3 is to the proof of existence of solutions to differential equations.

Theorem 5.6 (Picard–Lindelöf Theorem). *Suppose we are given the differential equation*

$$\frac{dy}{dx} = f(x, y) \text{ with initial condition } y(x_0) = y_0. \tag{5.53}$$

We may assume $x_0 = y_0 = 0$ (by taking appropriate translations if necessary). Suppose f is continuous in a rectangle $D = \{(x, y): |x| \le a, |y| \le b\}$ where a and b are positive constants. Also suppose that f satisfies a Lipschitz condition in y in D, namely that there exists a positive constant k such that

$$|f(x, y_1) - f(x, y_2)| \le k|y_1 - y_2| \quad \text{for all } (x, y_1) \text{ and } (x, y_2) \in D. \tag{5.54}$$

Then there exists a constant $\alpha > 0$ so that in the interval $I = \{x: |x| \le \alpha\}$ there exists a unique solution $y = \phi(x)$ to the initial-value problem (5.53).

Proof. Before giving the proof we give some motivation. Suppose we already knew $\phi(x)$ exists. Then

$$\phi(x) = \int_0^x f(t, \phi(t)) \, dt \quad \text{for } x \in I \tag{5.55}$$

by Theorem 2.5. Let M be the space of continuous functions on the closed interval I. M is a complete metric space with metric

$$d(f, g) = \max_{x \in I} |f(x) - g(x)|.$$

Define F from M to itself as follows: if $\psi \in M$, define $F(\psi)$ by

$$F(\psi)(x) = \int_0^x f(t, \psi(t)) \, dt \qquad (x \in I).$$

So if $\phi(x)$ exists, then it is a fixed point of F; i.e., $F(\phi) = \phi$. By Theorem 4.3, F will have a unique fixed point if it is a contraction. Now let $\phi_1, \phi_2 \in M$. Suppose in addition that

$$|\phi_1(t)| \le b \quad \text{and} \quad |\phi_2(t)| \le b \quad \text{for all } t \in I, \tag{5.56}$$

$$d(F(\phi_1), F(\phi_2)) = \max_{x \in I} |F(\phi_1)(x) - F(\phi_2)(x)|$$

$$= \max_{x \in I} \left| \int_0^x f(t, \phi_1(t)) \, dt - \int_0^x f(t, \phi_2(t)) \, dt \right|$$

$$\leq \max_{t,x\in I} |f(t,\phi_1(t)) - f(t,\phi_2(t))| \, |x - 0| \qquad (5.57)$$

$$\leq k \max_{t\in I} |\phi_1(t) - \phi_2(t)| \, \alpha \qquad (5.58)$$

$$= \alpha k \, d(\phi_1, \phi_2). \qquad (5.59)$$

We will want to choose α so that $\alpha k < 1$ in order that F be a contraction. Also we will want for any $\phi_1, \phi_2 \in M$ that (5.56) be satisfied so that (5.54) will allow us to deduce (5.58) from (5.57). We are now ready to prove the theorem. Since $f(x,y)$ is continuous on D there is a positive constant Q such that $|f(x,y)| \leq Q$ for all $(x,y) \in D$. Choose α sufficiently small so that $\alpha k = \lambda < 1$ and $\alpha < b/Q$. We now define $I = \{x: |x| \leq \alpha\}$ and modify our original definition of M so that

$$M = \{\phi: \phi \in C(I) \text{ and } |\phi(t)| \leq b \text{ for all } t \in I\},$$

M is a complete metric space. To see that F maps M into M, for $\phi \in M$ and $x \in I$,

$$|\phi(x)| = \left| \int_0^x f(t,\phi(t)) \, dt \right| \leq \alpha Q < b.$$

The derivation of the inequality (5.59) is now valid. Since $\alpha k < 1$, F is a contraction on M and therefore has a fixed point ϕ. We may choose $\phi_0(x)$ as the (constant) zero function, and for all i perform Picard iteration

$$\phi_{i+1}(x) = F(\phi_i)(x) = \int_0^x f(t,\phi_i(t)) \, dt \text{ for all } x \in I.$$

Then $\{\phi_i(x)\} \to \phi(x)$ as $i \to \infty$.

With one minor difference, we have given the same proof twice. In finding a solution to (5.50), F is a contraction when λ is sufficiently small; in finding a solution to (5.55), F is a contraction when the interval of integration from 0 to x is sufficiently small. $\qquad \square$

Recall from our discussion of the mean value theorem for derivatives that if $f: \mathbb{R} \to \mathbb{R} \in C^{(1)}$ and $x, x + \Delta x \in (a,b)$, then

$$f(x + \Delta x) = f(x) + f'(\eta) \Delta x \qquad (5.60)$$

for some η between x and $x + \Delta x$. However, if $f: \mathbb{R}^n \to \mathbb{R}^n \in C^{(1)}$ so that

$$f(\mathbf{x}) = \begin{pmatrix} f_1(\mathbf{x}) \\ \vdots \\ f_n(\mathbf{x}) \end{pmatrix}, \quad f'(\mathbf{x}) = \begin{pmatrix} \dfrac{\partial f_1}{\partial x_1} & \cdots & \dfrac{\partial f_1}{\partial x_n} \\ \vdots & \ddots & \vdots \\ \dfrac{\partial f_n}{\partial x_1} & \cdots & \dfrac{\partial f_n}{\partial x_n} \end{pmatrix},$$

then it is not true in general that given $\mathbf{x}, \Delta\mathbf{x} \in \mathbb{R}^n$ there exists η between (in the line segment joining) $\mathbf{x}$ and $\mathbf{x} + \Delta\mathbf{x}$ such that (5.60) holds. An example [18] is

$$f: \mathbb{R}^2 \to \mathbb{R}^2, \quad f\begin{pmatrix} x_1 \\ x_2 \end{pmatrix} = \begin{pmatrix} e^{x_1} - x_2 \\ x_1^2 - 2x_2 \end{pmatrix}.$$

Our concern is as follows. For $f: [a, b] \to [a, b]$ to be a contraction in $[a, b]$, if $|f'(x)| \le \lambda < 1$ on $[a, b]$, then f is a contraction on $[a, b]$ by Corollary 2.4:

$$|f(x) - f(y)| = |f'(\eta)(x - y)| \le \lambda |x - y|$$

for some η. We cannot extend this argument directly to $f: \mathbb{R}^n \to \mathbb{R}^n$. However

$$f(\mathbf{x} + \Delta\mathbf{x}) - f(\mathbf{x}) = \int_{\mathbf{x}}^{\mathbf{x}+\Delta\mathbf{x}} f'(\mathbf{z})\, d\mathbf{z} = \int_0^1 f'(\mathbf{x} + t\Delta\mathbf{x})\,\Delta\mathbf{x}\, dt$$

by componentwise application of Theorem 2.5. This implies that

$$\|f(\mathbf{x} + \Delta\mathbf{x}) - f(\mathbf{x})\| \le M \,\|\Delta\mathbf{x}\| \quad \text{where} \quad M = \max_{\mathbf{x} \in \bar{U}} \|f'(\mathbf{x})\|,$$

where $\bar{U}$ is a closed neighborhood containing $\mathbf{x}$ and $\mathbf{x}+\Delta\mathbf{x}$. Using this, we can prove the following important theorem:

Theorem 5.7 (Implicit Function Theorem). *Let $F \in C^{(1)}(U \times V, W)$ where U, V, W are open subsets of $\mathbb{R}^n$, $\mathbb{R}^m$, $\mathbb{R}^n$, respectively. Let $(\mathbf{x}_0, \mathbf{y}_0) \in U \times V$ with $F(\mathbf{x}_0, \mathbf{y}_0) = \mathbf{0}$ and*

$$D_{\mathbf{x}}F(\mathbf{x}_0, \mathbf{y}_0) = \begin{pmatrix} \dfrac{\partial F_1}{\partial x_1} & \cdots & \dfrac{\partial F_1}{\partial x_n} \\ \vdots & \ddots & \vdots \\ \dfrac{\partial F_n}{\partial x_1} & \cdots & \dfrac{\partial F_n}{\partial x_n} \end{pmatrix}(\mathbf{x}_0, \mathbf{y}_0)$$

be nonsingular. Then there exists a neighborhood $U_1 \times V_1 \subset U \times V$ and a function $f: V_1 \to U_1$, $f \in C^{(1)}$, where $f(\mathbf{y}_0) = \mathbf{x}_0$ such that

$$F(\mathbf{x}, \mathbf{y}) = \mathbf{0} \ \text{ for } (\mathbf{x}, \mathbf{y}) \in U_1 \times V_1 \iff \mathbf{x} = f(\mathbf{y}).$$

Proof. The basic idea is that if we have more unknowns than equations, we may choose and rename surplus variables as $y_1, \ldots, y_m$. Then, holding these variables constant, we may solve the set of equations in suitable neighborhoods for $x_1, \ldots, x_n$ by Newton's method. Staying nearby, as the values of $y_1, \ldots, y_m$ vary, so will the values of $x_1, \ldots, x_n$. Usually we do not know an explicit formula, but $x_1, \ldots, x_n$ are determined implicitly by $y_1, \ldots, y_m$. In the above, we use the standard notation

$$U \times V = \{(x, y): x \in U \text{ and } y \in V\}.$$

We now give the proof. Since $D_{\mathbf{x}}F(\mathbf{x}_0, \mathbf{y}_0)$ is nonsingular, there is a neighborhood of $(\mathbf{x}_0, \mathbf{y}_0)$ in which $D_{\mathbf{x}}F$ is nonsingular (the determinant is a continuous function that is nonzero at a point, and hence in a neighborhood). By choosing U and V smaller if necessary, $(D_{\mathbf{x}}F(\mathbf{x}, \mathbf{y}))^{-1}$ is defined on $U \times V$. Define $G: U \times V \to \mathbb{R}^n$ by

$$G(\mathbf{x}, \mathbf{y}) = \mathbf{x} - (D_{\mathbf{x}}F(\mathbf{x}, \mathbf{y}))^{-1} F(\mathbf{x}, \mathbf{y}).$$

Then $G(\mathbf{x}_0, \mathbf{y}_0) = \mathbf{x}_0$ and $D_{\mathbf{x}}G(\mathbf{x}_0, \mathbf{y}_0) = I - I = 0$, where I is the identity matrix. Since $G(\mathbf{x}_0, \mathbf{y}_0) = \mathbf{x}_0$ and $D_{\mathbf{x}}G(\mathbf{x}_0, \mathbf{y}_0) = 0$, and since $G \in C^{(1)}$, there exists a neighborhood $U_1 \times V_1$ of $(\mathbf{x}_0, \mathbf{y}_0)$ with $\|D_x G(\mathbf{x}, \mathbf{y})\| \leq \alpha < 1$ in $\bar{U}_1 \times \bar{V}_1$. Choose this neighborhood $U_1 \times V_1$ such that $G \colon \bar{U}_1 \times \bar{V}_1 \to \bar{U}_1$ (see above). For each $\mathbf{y} \in V_1$, $G(\mathbf{x}, \mathbf{y}) \colon \bar{U}_1 \to \bar{U}_1$ is a contraction, and hence has a unique fixed point which we denote by $f(\mathbf{y})$. To see that f is smooth, and to see an illustrative example, consult Edwards [23].   □

A special case of the preceding proof guarantees convergence of Newton's method (under reasonable hypotheses) if the initial guess is sufficiently close to the solution.

Theorem 5.8. *Let $F \in C^{(1)}(U, W)$ with U, W open in $\mathbb{R}^n$. Suppose $F(\xi) = 0$ and $F'(\xi)$ is nonsingular. Then there is a neighborhood U_1 of ξ such that if $\mathbf{x}_0 \in U_1$ and*

$$\mathbf{x}_{n+1} = \mathbf{x}_n - (F'(\mathbf{x}_n))^{-1} F(\mathbf{x}_n)$$

for all n, then the sequence $\{\mathbf{x}_n\}$ converges to ξ.

Proof. Consider $m = 0$ and $\mathbb{R}^m$ as the empty set. That is, take $F(\mathbf{x})$ instead of $F(\mathbf{x}, \mathbf{y})$. $G \colon U \to \mathbb{R}^n$ becomes $G(\mathbf{x}) = \mathbf{x} - (F'(\mathbf{x}))^{-1} F(\mathbf{x})$ on sufficiently small U_1, with $G \colon \bar{U}_1 \to \bar{U}_1$ a contraction; hence if $\mathbf{x}_0 \in U_1$, then $\{\mathbf{x}_{n+1} = G(\mathbf{x}_n)\}$ converges to the fixed point ξ of G (where $F(\xi) = 0$).   □

For a treatment of the implicit function theorem, existence of solutions of differential equations, and related topics in greater generality, see Chow and Hale [15].

5.17 A Duality Theorem and Cost Minimization

The proof of the duality theorem and an example are by Duffin [20, 21]. Suppose $c_1, \ldots, c_n$ is a sequence of positive constants and for each $i = 1, \ldots, n$ there is a sequence of real numbers $\alpha_{i1}, \ldots, \alpha_{ik}$. Suppose these are used to define for positive $t_1, \ldots, t_k$ the cost function

$$u(t_1, \ldots, t_k) = c_1 t_1^{\alpha_{11}} \cdots t_k^{\alpha_{1k}} + \cdots + c_n t_1^{\alpha_{n1}} \cdots t_k^{\alpha_{nk}}.$$

Denote

$$\mathbb{R}_k^+ = \{\mathbf{t} = (t_1, \ldots, t_k) \colon \text{each } t_i > 0\},$$

$$\Delta_n = \left\{ \boldsymbol{\delta} = (\delta_1, \ldots, \delta_n) \colon \text{each } \delta_i > 0 \text{ and } \sum_{i=1}^{n} \delta_i = 1 \right\},$$

$$\Delta_n^\alpha = \left\{ \boldsymbol{\delta} \colon \boldsymbol{\delta} \in \Delta_n \text{ and } \sum_{i=1}^{n} \alpha_{ij} \delta_i = 0 \text{ for } j = 1, \ldots, k \right\},$$

$$P_i(\mathbf{t}) = t_1^{\alpha_{i1}} \cdots t_k^{\alpha_{ik}} \quad \text{for } \mathbf{t} \in \mathbb{R}_k^+,$$

$$v(\boldsymbol{\delta}) = \prod_{i=1}^{n} \left(\frac{c_i}{\delta_i} \right)^{\delta_i} \quad \text{for } \boldsymbol{\delta} \in \Delta_n.$$

Using the above notation with the positive constants $c_1, \ldots, c_n$ and the matrix of real numbers (the exponents in the cost function) $\{\alpha_{ij}\}$ fixed, we state two problems:

Problem 1: Let $M = \inf_{\mathbf{t} \in \mathbb{R}_k^+} \{u(\mathbf{t})\}$. Find M.

Problem 2 (The dual problem): Let $m = \sup_{\delta \in \Delta_n^\alpha} \{v(\delta)\}$. Find m.

Zero is clearly a lower bound for the set in Problem 1 hence M exists. On the interval $(0, 1)$, $(1/\delta_i)^{\delta_i}$ is bounded by $e^{1/e}$ so m exists for Problem 2.

To show how the two problems are related we apply the weighted AM–GM inequality to $u(\mathbf{t})$: for any $\delta \in \Delta_n$ and $\mathbf{t} \in \mathbb{R}_k^+$,

$$u(\mathbf{t}) = \sum_{i=1}^n \delta_i \left(\frac{c_i P_i(\mathbf{t})}{\delta_i} \right) \geq \prod_{i=1}^n \left(\frac{c_i P_i(\mathbf{t})}{\delta_i} \right)^{\delta_i} = v(\delta)\, t_1^{D_1} \cdots t_k^{D_k},$$

where each

$$D_j = \sum_{i=1}^n \alpha_{ij} \delta_i.$$

If $\delta \in \Delta_n^\alpha$, then each $D_j = 0$. Thus $u(\mathbf{t}) \geq v(\delta)$ for all $\mathbf{t} \in \mathbb{R}_k^+$ and $\delta \in \Delta_n^\alpha$. For $\mathbf{t} \in \mathbb{R}_k^+$ and $\delta \in \Delta_n$ define

$$Q(\mathbf{t}, \delta) = u(\mathbf{t}) - v(\delta)\, t_1^{D_1} \cdots t_k^{D_k}.$$

Then $Q(\mathbf{t}, \delta) \geq 0$ with equality if and only if all $c_i P_i(\mathbf{t})/\delta_i$ are equal. Now suppose that $u(\mathbf{t})$ attains its infimum M at a point $\mathbf{t}^* \in \mathbb{R}_k^+$. For each i choose $\delta_i^* = c_i P_i(\mathbf{t}^*)/M$. Then $\delta^* \in \Delta_n$ and all $c_i P_i(\mathbf{t}^*)/\delta_i^*$ are equal, hence $Q(\mathbf{t}^*, \delta^*) = 0$. Since we have assumed the cost function $u(\mathbf{t})$ attains its minimum at $\mathbf{t}^*$ in the open set $\mathbb{R}_k^+$, $\partial u/\partial t_j = 0$ for $j = 1, \ldots, k$ at this point. Since $Q(\mathbf{t}, \delta)$ attains its minimum at $(\mathbf{t}^*, \delta^*)$, all its partial derivatives, hence the first k, $\partial Q/\partial t_j = 0$ at $(\mathbf{t}^*, \delta^*)$. But

$$\frac{\partial Q}{\partial t_j} = \frac{\partial}{\partial t_j} u - \frac{\partial}{\partial t_j} v(\delta)\, t_1^{D_1} \cdots t_k^{D_k}.$$

Since the first term is zero,

$$\frac{\partial}{\partial t_j} v(\delta)\, t_1^{D_1} \cdots t_k^{D_k} = D_j v(\delta)\, t_1^{D_1} \cdots t_j^{D_j - 1} \cdots t_k^{D_k} = 0.$$

The conditions $D_j = 0$ are exactly that $\delta \in \Delta_n^\alpha$. Thus $\delta^* \in \Delta_n^\alpha$ and $v(\delta^*) = M$. Since $v(\delta) \leq u(\mathbf{t})$ for all $\mathbf{t} \in \mathbb{R}_k^+$ and $\delta \in \Delta_n^\alpha$, in particular $v(\delta) \leq u(\mathbf{t}^*) = M$ for all $\delta \in \Delta_n^\alpha$ with equality when $\delta = \delta^*$. Thus $M = m$ and

$$v(\delta) \leq M \leq u(\mathbf{t}) \quad \text{for all } \delta \in \Delta_n^\alpha \text{ and } \mathbf{t} \in \mathbb{R}_k^+ \tag{5.61}$$

with equality at $\mathbf{t} = \mathbf{t}^*$ and $\delta = \delta^*$. Therefore the following has been proved:

Theorem 5.9. *If the cost function $u(\mathbf{t})$ in Problem 1 attains its infimum M in $\mathbb{R}_k^+$, then the dual function $v(\delta)$ in Problem 2 has maximum value M in Δ_n^α.*

Example 5.29. Suppose $400 \, \text{yd}^3$ of material must be ferried across a river in an open box of length t_1, width t_2, and height t_3. The bottom and sides cost $10 per yd^2 and the ends cost $20 per yd^2. Two runners costing $2.50 per yd are needed for the box to slide on. Each round trip of the ferry costs $0.10. Minimize the total cost

$$u(t_1, t_2, t_3) = \frac{40}{t_1 t_2 t_3} + 20 t_1 t_3 + 40 t_2 t_3 + 10 t_1 t_2 + 5 t_1.$$

(Ignore the fact that a fraction of a trip does not make sense.)

The reader might wish to perform the following numerical experiment. Apply Newton's method to the gradient of $u(\mathbf{t})$, hoping to find a minimum point. In order to force the constraints that $t_i > 0$, substitute $t_i = \varepsilon + x_i^2$ and estimate the minimum of $u(\mathbf{x})$ by applying Newton's method to its gradient. (When using Newton's method, use the Jacobian derivative of the gradient of u, which is the Hessian of u.) Set $\varepsilon = 0$, take initial guesses $t_i = x_i^2 = 1$, and find values of $t_1 \approx 1.54$, $t_2 \approx 1.11$, $t_3 \approx 0.557$. Take this value of $\mathbf{t}$ as (a good approximation for) $\mathbf{t}^*$ and define $\delta_i^* = c_i P_i(\mathbf{t}^*)/u(\mathbf{t}^*)$ for all i and verify (to within a specified tolerance) that $\delta^* \in \Delta_n^\alpha$ and $u(\mathbf{t}^*) = v(\delta^*) = 108.69$. Note that a typical application of Newton's method yields only a local result. However, (5.61) tells us that we found $M = \$108.69$, the global minimum cost.   □

5.18 Problems

5.1. Use the Cauchy–Schwarz inequality to obtain upper bounds for the integrals

(a) $I_1 = \int_0^1 \sqrt{1 + x^3} \, dx$,

(b) $I_2 = \int_0^\pi \sqrt{x \sin x} \, dx$,

(c) $I_3 = \int_0^\pi \sqrt{\sin x} \, dx$.

5.2. Use Chebyshev's inequality to

(a)   obtain an upper bound for the integral

$$I = \int_2^5 \frac{e^x}{x + 1} \, dx,$$

(b)   derive the inequality

$$(\sin^{-1} t)^2 < \frac{t}{2} \ln \left| \frac{t + 1}{t - 1} \right|.$$

5.3. Use the Darboux inequality (2.8) to find an upper bound for the complex contour integral

$$I = \int_C \frac{dz}{z^2 + a^2}$$

where C is the contour $z = b e^{j\theta}$ for $0 \leq \theta \leq \theta_0$ $(b \neq a)$.

5.4. Show that if $u(x)$ and every $u_n(x)$ are integrable on $[a, b]$, and if $\{u_n(x)\}$ converges uniformly to $u(x)$ on $[a, b]$, then

$$\int_a^b \lim_{n \to \infty} u_n(x)\, dx = \int_a^b u(x)\, dx = \lim_{n \to \infty} \int_a^b u_n(x)\, dx.$$

5.5. Let $u(x)$ and the sequence $\{u_n(x)\}$ be defined on $[a, b]$. We say that $\{u_n(x)\}$ converges in the *mean square sense* to $u(x)$ if and only if

$$\lim_{n \to \infty} \int_a^b [u(x) - u_n(x)]^2\, dx = 0.$$

(a)   Show that uniform convergence implies mean square convergence.

(b)   Show that mean square convergence implies

$$\lim_{n \to \infty} \int_a^b u_n^2(x)\, dx = \int_a^b u^2(x)\, dx.$$

5.6. Write a computer program using Simpson's rule to estimate $\int_0^1 e^{x^2}\, dx$. Keep doubling the number of partitions until convergence within a specified tolerance. Do not recompute any function evaluations.

5.7. Show that

$$2\Gamma(n + \tfrac{1}{2}) \le \Gamma(\tfrac{1}{2})\Gamma(n + 1) \le 2^n \Gamma(n + \tfrac{1}{2}) \qquad (n \in \mathbb{N})$$

with strict inequality for $n > 1$. See [50] for an application to traffic flow.

5.8. For the exponential integrals defined in the text, show that $E_{n+1}(x) < E_n(x)$ for $n \in \mathbb{N}$.

5.9. Show that for $x > 1$

$$\frac{x - 1}{x^2} e^{-x} < \int_x^\infty \frac{e^{-t}}{t}\, dt < \frac{e^{-x}}{x}.$$

5.10. The *autocorrelation* of a real-valued function f is the function $f \star f$ defined by

$$f \star f(t) = \int_{-\infty}^\infty f(u)f(t + u)\, du.$$

Show that $f \star f$ reaches its maximum at $t = 0$.

5.11. Use the power series expansion for $J_n(x)$ to show that for $n \ge 0$,

$$|J_n(x)| \le \frac{|x|^n}{2^n (n!)} e^{x^2/4}.$$

5.12. The complementary error function $\operatorname{erfc}(x)$ is defined by

$$\operatorname{erfc}(x) = \frac{2}{\sqrt{\pi}} \int_x^\infty e^{-t^2}\, dt.$$

(a)   Establish the following upper bound:

$$\operatorname{erfc}(\sqrt{x}) < \frac{e^{-x}}{\sqrt{\pi x}}.$$

This bound is useful for $x > 3$, but obviously not for small x.

(b)   Show that for $0 \le x \le y$ we have $\mathrm{erfc}(\sqrt{y}) \le e^{-(y-x)}\,\mathrm{erfc}(\sqrt{x})$.

5.13. (See Anderson et al. [3]). Let a, b be real, positive numbers. Denote the arithmetic and geometric means by $A(a, b) = (a + b)/2$ and $G(a, b) = \sqrt{ab}$, respectively. Define sequences starting with $a_0 = a$, $b_0 = b$ and, for all n, $a_{n+1} = A(a_n, b_n)$, $b_{n+1} = G(a_n, b_n)$.

(a)   Use AM–GM and induction to show that $b_n \le b_{n+1} \le a_{n+1} \le a_n$ for $n \ge 1$.

(b)   Observe that a_n is a decreasing sequence bounded below (by b_0), and hence has a limit. Similarly, b_n has a limit.

(c)   Show that the sequences a_n and b_n have a common limit. This common limit, the *arithmetic–geometric mean* of a and b, is denoted by $AG(a, b)$.

(d)   Defining the integral

$$T(a, b) = \int_0^{\pi/2} \frac{dx}{\sqrt{a^2 \cos^2 x + b^2 \sin^2 x}}$$

with a_1 and b_1 defined as above, show that $T(a, b) = T(a_1, b_1)$.

(e)   Show that the sequence $\sqrt{a_n^2 \cos^2 x + b_n^2 \sin^2 x}$ converges uniformly to $AG(a, b)$ on $[0, \pi/2]$.

(f)   Show that

$$T(a, b) = \frac{\pi/2}{AG(a, b)}.$$

(g)   Now let $0 < r < 1$ and set $a = 1$ and $b = \sqrt{1 - r^2}$ to deduce the stunning result of Gauss concerning Legendre's elliptic integrals of the first kind:

$$\int_0^{\pi/2} \frac{dx}{\sqrt{1 - r^2 \sin^2 x}} = \frac{\pi/2}{AG(1, \sqrt{1 - r^2})}.$$

(h)   Let $r = 1/2$. Verify that you get six digits of accuracy in using $(\pi/2)/a_2$ to estimate the elliptic integral

$$\int_0^{\pi/2} (\sqrt{1 - r^2 \sin^2 x})^{-1} dx$$

5.14. Show that if $\nabla^2 \Phi = 0$ throughout a region bounded by a simple closed surface S, then

$$\oint_S \Phi \frac{\partial \Phi}{\partial n} dS \ge 0$$

5.15. Two isolated conducting bodies carry electric charges. Show that if they are subsequently connected by a very thin wire, the total stored energy of the system is diminished.

5.16. Let $A = \begin{pmatrix} 1 & 1 \\ 0 & 1 \end{pmatrix}$. Show that there is no $X \ne 0$ with $\|AX\|_2 = \|A\|_F \|X\|_2$.

5.17. Show that of all finite-energy signal forms, the Gaussian pulse $f(t) = K_2 \exp(K_1 t^2/2)$, where K_1, K_2 are constants with $K_1 < 0$, have the minimum duration-bandwidth product.

5.18. Show that an improved lower bound on the coerror function is given for $x > 0$ by

$$Q(x) > \frac{1}{\sqrt{2\pi}} \frac{x}{x^2 + 1} e^{-x^2/2}.$$

5.19. Use probability concepts to derive the bound $Q(x) \le \frac{1}{2} e^{-x^2/2}$ for $x \ge 0$.

5.20. [46] Let X be a Poisson random variable with mean λ and develop the Chernoff bound

$$P(X \geq k) \leq \frac{(\lambda e)^k}{k^k} e^{-\lambda} \qquad (k > \lambda)$$

5.21. Use Picard iteration to solve $y' = 2y$ subject to $y(0) = 1$.

5.22. Describe a suitable choice of the neighborhood U_1 when using Newton's method (Theorem 5.8) for the case $n = 1$.

Chapter 6
Inequalities for Differential Equations

In the theory of differential equations, inequalities are widely used to estimate or approximate solutions to problems. They are also needed to establish uniqueness and existence, along with other theoretical results pertaining to solution behavior. The purpose of this chapter is to touch on a few inequalities that play key roles in the study of differential equations.

6.1 Chaplygin's Theorem

Following Chaplygin [14], we consider the relation between the solutions of a first-order ordinary differential equation and a certain inequality.

Theorem 6.1. *Assume that for $t > a$ the functions y and z satisfy*

$$y'(t) - f(t, y(t)) = 0 , \qquad z'(t) - f(t, z(t)) > 0 , \tag{6.1}$$

and that $y(a) = z(a)$. Then $z(t) > y(t)$ for $t > a$.

Proof. Combining the relations (6.1), we get

$$z'(t) - y'(t) - p(t)[z(t) - y(t)] > 0 \quad (t > a) \quad \text{where } p(t) = \frac{f(t, z(t)) - f(t, y(t))}{z(t) - y(t)} .$$

The function p is uniquely defined through the unknown y, z. Let us multiply both sides of the inequality by $\exp\left[-\int_a^t p(s)\,ds\right]$. The result can be written as

$$\frac{d}{dt}\left\{[z(t) - y(t)] \exp\left[-\int_a^t p(s)\,ds\right]\right\} > 0 .$$

M.J. Cloud et al., *Inequalities: With Applications to Engineering*,
DOI 10.1007/978-3-319-05311-0_6, © Springer International Publishing AG 2014

Integrating the inequality with respect to t between a and τ, we get

$$[z(t) - y(t)] \exp\left[-\int_a^t p(s)\,ds\right]\Big|_a^\tau > 0 \,.$$

Because $z(a) = y(a)$ and the exponential is positive, we have $z(\tau) > y(\tau)$.   □

Similarly, it can be shown that if there is function x such that $x'(t) - f(t, x(t)) < 0$ and $x(a) = y(a)$, then for all $t > a$ we have a lower bound for y as $y(t) > x(t)$.

In [14] the result is extended to the solutions of an equation and inequality of the nth order:

$$y^{(n)} - f(t, y, \ldots, y^{(n-1)}) = 0 \,, \quad z^{(n)} - f(t, z, \ldots, z^{(n-1)}) > 0 \qquad (t > a)$$

with coinciding initial values

$$y(a) = z(a) \,, \quad \ldots \,, \quad y^{(n-1)}(a) = z^{(n-1)}(a) \,.$$

Here we have $z(t) > y(t)$ as well.

An approach based on similar ideas is presented next.

6.2 Gronwall's Inequality

Also known as *Gronwall's lemma*, the main result of this section is widely used in the theory of ordinary and dynamic partial differential equations. We formulate it (see, e.g., [35]) as follows.

Theorem 6.2 (Gronwall's Inequality). *Let $z(t)\colon [a, b] \rightarrow \mathbb{R}$ be a nonnegative continuous function satisfying the following inequality on $[a, b]$:*

$$z(t) \le C + \int_a^t p(s)z(s)\,ds \,, \tag{6.2}$$

where p is a nonnegative continuous function on $[a, b]$ and the constant $C \ge 0$. Then on $[a, b]$ the function z satisfies

$$z(t) \le C \exp\left[\int_a^t p(s)\,ds\right] \,. \tag{6.3}$$

Proof. First we suppose that $C > 0$. Denote

$$Z(t) = C + \int_a^t p(s)z(s)\,ds$$

so (6.2) is $z(t) \leq Z(t)$. As $z(t) \geq 0$ and $p(t) \geq 0$, we also have $Z(t) \geq C > 0$. Differentiating Z, we get $Z'(t) = p(t)z(t)$ and as $p(t) \geq 0$ we deduce that $Z'(t) \leq p(t)Z(t)$. This is equivalent to the inequality

$$\frac{Z'(t)}{Z(t)} \leq p(t) , \qquad Z(a) = C .$$

Integrating both sides, we get

$$\ln Z(t) - \ln Z(a) \leq \int_a^t p(s)\,ds$$

which implies (6.3).

Finally, when $C = 0$ we can use the established result (6.3). Taking in (6.3) a sequence of constants $C_k \to +0$ as $k \to \infty$, by continuity of p we obtain $z(t) = 0$. $\square$

Note that Gronwall's lemma can be derived from Chaplygin's theorem.

Example 6.1. We will use Theorem 6.2 to establish a uniqueness theorem for the linear system of equations

$$\mathbf{x}'(t) = A(t)\mathbf{x}(t) + \mathbf{f}(t)$$

with initial condition

$$\mathbf{x}(a) = \mathbf{x}_0 ,$$

where the column vector $\mathbf{x}(t) = (x_1(t), \ldots, x_n(t))^T$, the matrix $A(t)$ is $n \times n$ with elements continuous for $t > a$, and the column vector $\mathbf{f}(t)$ is a given vector function with piecewise continuous components.

Suppose there are two solutions to the problem:

$$\mathbf{x}_k'(t) = A(t)\mathbf{x}_k(t) + \mathbf{f}(t) , \qquad \mathbf{x}_k(a) = \mathbf{x}_0 \qquad (k = 1, 2) .$$

Subtracting the equalities and denoting $\mathbf{x}(t) = \mathbf{x}_2(t) - \mathbf{x}_1(t)$, we get

$$\mathbf{x}'(t) = A(t)\mathbf{x}(t) , \qquad \mathbf{x}(a) = \mathbf{0} .$$

Let us integrate the last equation with respect to t over $[a, \tau]$:

$$\mathbf{x}(\tau) = \int_a^\tau A(t)\mathbf{x}(t)\,dt .$$

Then

$$\|\mathbf{x}(\tau)\| = \left\| \int_a^\tau A(t)\mathbf{x}(t)\,dt \right\| \leq \int_a^\tau \|A(t)\|\,\|\mathbf{x}(t)\|\,dt ,$$

which is a particular case of (6.2) with $C = 0$. Thus $\mathbf{x}(t) = \mathbf{0}$ as needed. $\square$

6.3 On Poisson's Equation

Poisson's equation

$$\nabla^2 u + f = 0 \tag{6.4}$$

describes the normal displacement u of a planar membrane S in equilibrium under a distributed force f. Although we encounter (6.4) in other applications (e.g., electromagnetics), its mechanical interpretation is most picturesque and allows us to imagine what happens during deformation of the membrane. To (6.4) we attach the boundary condition that the edge of the membrane ∂S is fixed:

$$u\big|_{\partial S} = 0 .$$

The membrane is elastic but does not obey Hooke's law. As for all elastic objects under conservative forces, we can describe membrane equilibrium using the principle of minimum total potential energy. So a sufficiently smooth solution to the above problem will minimize the energy integral

$$E(u) = \frac{1}{2} \int_S (u_x^2 + u_y^2)\, dx\, dy - \int_S fu\, dx\, dy$$

on the set of twice-differentiable functions that vanish on ∂S. We refer to these functions as *admissible displacements*. The minimum energy principle can be established via the calculus of variations, which falls outside the scope of this book (the interested reader could see, e.g., [47]). However, we should say that the minimizer would satisfy the equation

$$\int_S (u_x v_x + u_y v_y)\, dx\, dy - \int_S fv\, dx\, dy = 0 \tag{6.5}$$

for any admissible displacement v (i.e., v is smooth and $v|_{\partial S} = 0$). This can be shown as follows. We assume u is a minimizer of E and so, considering $E(u+tv)$ with fixed u, v as a function of the real variable t, obtain a function that should take its minimum value at $t = 0$. Differentiation gives us (6.5), which must hold for any admissible v. This equality is the basis for introducing weak solutions to the membrane problem. As the bilinear form

$$\int_S (u_x v_x + u_y v_y)\, dx\, dy \tag{6.6}$$

turns out to be an inner product, we are immediately led to apply Hilbert space methods to the problem. To this end, we must examine certain integral relations between v appearing in the integral

$$\int_S fv\, dx\, dy \tag{6.7}$$

and its first derivatives. Unfortunately, the admissible functions fail to constitute a complete space under the inner product (6.6). So to carry out the plan we must extend the space to a Hilbert space. This process entails topics such as Lebesgue integration, generalized derivatives, and Sobolev spaces, which fall outside our scope as well. These topics are important in mechanics and other areas of mathematical physics, and at some point the reader may take up their study. At this stage we merely add that the resulting collection of topics underlies the theory of Ritz's method and its offshoots such as the finite element method. So here we will consider some inequalities involved in this theory, selecting points appropriate to the level of this book.

6.4 Membrane with Fixed Edge

Let us apply Schwarz's inequality to the integral (6.7):

$$\left| \int_S fv\,dx\,dy \right| \le \left(\int_S f^2\,dx\,dy \right)^{1/2} \left(\int_S v^2\,dx\,dy \right)^{1/2}.$$

If we assume f is square-integrable, then an estimate of the last integral

$$\int_S v^2\,dx\,dy$$

in terms of the strain energy

$$D(v) = \int_S (v_x^2 + v_y^2)\,dx\,dy$$

can be quite useful.

We will prove that there is a constant C, depending only on S, such that

$$\int_S v^2\,dx\,dy \le C \int_S (v_x^2 + v_y^2)\,dx\,dy \tag{6.8}$$

for all admissible functions v. This is the *Friedrichs inequality* (recall the one-dimensional case stated in Theorem 3.10).

Note that, when applying this estimate in the theory, we typically do not need the *best* value for C and will not try to obtain it. However, some literature is devoted to this question as the best value does play an important role in the theory of oscillation of the homogeneous membrane. We also note that (6.8) may be regarded as one of the first results hailing the appearance of the Sobolev spaces and their associated imbedding theorems, an important aspect of modern mathematical physics.

We prove (6.8) for a bounded domain S having piecewise smooth boundary ∂S. Let S be a portion of the rectangle

$$R = \{(x, y): a < x < b, \ c < y < d\}.$$

We recall that

$$v\big|_{\partial S} = 0 .$$ (6.9)

Let us extend v by zero outside of S. Now v is continuous but its first derivatives are merely piecewise continuous as they can have jump discontinuities at ∂S. However, since $v(a, y) = 0$ for any y, we can apply the representation

$$v(x, y) = \int_a^x v_s(s, y) \, ds .$$

Squaring this and applying Schwarz's inequality, we get

$$v^2(x, y) = \left(\int_a^x 1 \cdot v_s(s, y) \, ds \right)^2 \le \int_a^x 1^2 \, ds \int_a^x v_s^2(s, y) \, ds$$

$$\le (b - a) \int_a^b v_s^2(s, y) \, ds .$$ (6.10)

Integration over R yields

$$\int_R v^2(x, y) \, dx \, dy \le (b - a) \int_R \int_a^b v_s^2(s, y) \, ds \, dx \, dy = (b - a)^2 \int_R v_x^2(x, y) \, dx \, dy .$$

Remembering that $v = 0$ and $v_x = 0$ outside S, we get

$$\int_S v^2(x, y) \, dx \, dy \le (b - a)^2 \int_S v_x^2(x, y) \, dx \, dy .$$

A similar inequality holds with v_x replaced by v_y. Summing these inequalities, we find that (6.8) holds with $C = \frac{1}{2}(b - a)^2$. The value of C can be reduced, but this is outside our scope. As we see, here c and d can be infinite, hence we should assume additionally that v_x is square-integrable over S.

We may extend (6.8) to functions continuously differentiable on a bounded domain $V \subset \mathbb{R}^n$ and vanishing on the boundary ∂V. This is done by mimicking the above transformations. First, extending v by zero outside V, we suppose the domain lies inside the band

$$\{\mathbf{x} = (x, \mathbf{y}) : a < x < b\}$$

and rewrite (6.10), substituting $\mathbf{y} \in \mathbb{R}^{n-1}$ for y:

$$v^2(x, \mathbf{y}) \le (b - a) \int_a^b v_s^2(s, \mathbf{y}) \, ds .$$

Integrating this over the band and then returning to V, we get

$$\int_V v^2(x, \mathbf{y}) \, dV \le (b - a)^2 \int_V v_x^2(x, \mathbf{y}) \, dV$$

from which it follows that

$$\int_V v^2 \, dV \le C \int_V \nabla v \cdot \nabla v \, dV \qquad (6.11)$$

for some constant C. To make this proof valid, we must include V in a band of finite width and rotate the coordinates so that x is the width coordinate.

Extensions of (6.8) may be found in the theory of Sobolev spaces, but the techniques involved in their proof are more difficult than those exhibited here. An inequality with some constant C independent of v holds if v is zero only on some part of ∂S and $1 \le p < \infty$:

$$\left(\int_S |v|^p \, dx \, dy \right)^{2/p} \le C_p \int_S (v_x^2 + v_y^2) \, dx \, dy \, ,$$

with constant C_p depending on bounded S and p only. Inequality (6.11) can be extended similarly, but the conditions on the boundary and p are more restrictive.

Finally, in Sobolev theory it is shown that from any sequence of smooth functions $\{v_n\}$ satisfying (6.9) and such that there is a constant c for which

$$\int_S (v_{nx}^2 + v_{ny}^2) \, dx \, dy \le c \, ,$$

we can select a subsequence $\{v_{n_k}\}$ that is Cauchy in the $L_2(S)$ norm. This property can be stated in terms of a *compact imbedding* from the Sobolev space $W^{1,2}(S)$ to the Lebesgue space $L_2(S)$ (see, e.g., [48]).

One of the inequalities used in mathematical physics is *Poincaré's inequality*. It relates the L_2 norm of a function similarly to (6.8), but holds without additional conditions on the boundary: there is a constant C such that for any bounded 2D domain S with piecewise smooth boundary we have for any continuously differentiable function u

$$\int_S u^2 \, dx \, dy \le C \left[\left(\int_S u \, dx \, dy \right)^2 + \int_S (u_x^2 + u_y^2) \, dx \, dy \right] . \qquad (6.12)$$

For a proof the reader can consult [25].

Now we extend inequality (6.8) to other applications.

6.5 Theory of Linear Elasticity

As we cannot explain in detail the theory of linear elasticity, we merely state that we will consider a deformable body occupying a bounded volume $V \subset \mathbb{R}^3$ with a piecewise smooth boundary ∂V. We denote the displacements of its points by a vector field $\mathbf{u}$ having components u_i. We assume these component functions are differentiable. Deformation of the body is defined by the strain tensor, which the reader may regard as a matrix (ε_{ij}) with i, j taking the values $1, 2, 3$. The fact that the ε_{ij} are components of a second-order tensor has many consequences, but these are not so important in our presentation as we will use only fixed Cartesian coordinates x_k. This same comment applies to all tensors mentioned below: the reader can regard them as some sets of constants or variable functions with indices. The quantities ε_{ij} are related to the displacement components by

$$\varepsilon_{ij} = \frac{1}{2}\left(\frac{\partial u_i}{\partial x_j} + \frac{\partial u_j}{\partial x_i}\right) \equiv \frac{1}{2}\left(u_{i,j} + u_{j,i}\right).$$

The notation $u_{i,j}$ means that we take the derivative of u_i with respect to x_j (note that j stands after the comma). In elasticity we also use the stress tensor with components σ_{ij} $(i, j = 1, 2, 3)$. The relation between (σ_{ij}) and (ε_{ij}) is the generalized form of Hooke's law

$$\sigma_{ij} = \sum_{k,m=1}^{3} c^{ijkm} \varepsilon_{km},$$

where the constants c^{ijkm} comprise the fourth-order *elasticity tensor*. The elastic properties of a homogeneous and isotropic body, however, can be expressed in terms of only two elastic constants: Young's modulus and Poisson's ratio. For present purposes, it is important only that c^{ijkm} possess the symmetry properties

$$c^{ijkm} = c^{jikm} = c^{kmij}$$

along with *positivity*: there is a positive constant m such that for any (ε_{ij}) we have

$$\sum_{i,j,k,m=1}^{3} c^{ijkm} \varepsilon_{km} \varepsilon_{ij} \geq m \sum_{k,m=1}^{3} \varepsilon_{km}^2. \tag{6.13}$$

The equilibrium equations in elasticity are written as

$$\sum_{m=1}^{3} \sigma_{km,m} + F_k = 0 \text{ in } V, \tag{6.14}$$

where the F_k are components of the distributed volume forces. If we wish to get the equilibrium equations in terms of displacements, we must substitute the above relations into (6.14). Elasticity problems are formulated by supplementing the equilibrium equations with boundary conditions. We will consider only the simplest one, which corresponds to the membrane with fixed edge:

$$\mathbf{u}\big|_{\partial V} = \mathbf{0}. \tag{6.15}$$

Now we recast the problem in weak form, as was done above for the membrane. We multiply the kth equation by v_k, an arbitrary function that vanishes on the boundary, sum over k, and integrate over V:

$$\int_V \sum_{k=1}^{3} \sum_{m=1}^{3} \sigma_{km,m} v_k \, dV + \int_V \sum_{k=1}^{3} F_k v_k \, dV = 0 \, .$$

A final transformation involves integration by parts in light of (6.15). Using the results above, we obtain

$$\int_V \sum_{i,j,k,m=1}^{3} c^{ijkm} \varepsilon_{km}(\mathbf{u}) \varepsilon_{ij}(\mathbf{v}) \, dV - \int_V \sum_{k=1}^{3} F_k v_k \, dV = 0 \, , \qquad (6.16)$$

where $\mathbf{v} = (v_1, v_2, v_3)$ and the symbols $\mathbf{u}, \mathbf{v}$ shown as arguments indicate which vector should have its components substituted into the components of the strain tensor. This equality must hold for any smooth v_k having zero values on the boundary. The "weak" setup (6.16) of the equilibrium problem, which is equivalent to (6.14)–(6.15), is also called the *virtual work principle of elasticity*. It is a consequence of the principle of minimum total potential energy of the body, which can be derived via the calculus of variations.

We will not discuss the question of existence of a weak solution to this problem. However, uniqueness of the smooth solution follows directly from (6.16). Suppose there are two solutions $\mathbf{u}_1$ and $\mathbf{u}_2$. Denote $\mathbf{u}^* = \mathbf{u}_2 - \mathbf{u}_1$ and take $\mathbf{v} = \mathbf{u}^*$. We get

$$\int_V \sum_{i,j,k,m=1}^{3} c^{ijkm} \varepsilon_{km}(\mathbf{u}^*) \varepsilon_{ij}(\mathbf{u}^*) \, dV = 0 \, ,$$

from which it follows that $\varepsilon_{ij}(\mathbf{u}^*) = 0$ for all i, j. It can be shown that, consequently,

$$\mathbf{u}^* = \mathbf{a} + \mathbf{x} \times \mathbf{b}$$

where $\mathbf{a}, \mathbf{b}$ are vector constants and $\mathbf{x}$ is the position vector of a point in the body. By (6.15) we have $\mathbf{u}^* = \mathbf{0}$.

Now we note that there are *plane problems* of elasticity. These are distinguished by the number of components that constitute the vectors and tensors involved. When each index can take only the values $1, 2$, the weak setup of the equilibrium problem is based on the equation

$$\int_S \sum_{i,j,k,m=1}^{2} c^{ijkm} \varepsilon_{km}(\mathbf{u}) \varepsilon_{ij}(\mathbf{v}) \, dS - \int_S \sum_{k=1}^{2} F_k v_k \, dS = 0 \, , \qquad (6.17)$$

where S is a 2D-domain and $\mathbf{u}, \mathbf{v}$ vanish on the boundary contour of S.

As for the membrane, the term containing the forces F_k can be estimated using Schwarz's inequality:

$$\left| \int_V \sum_{k=1}^{3} F_k v_k \, dV \right| \leq \sum_{k=1}^{3} \left(\int_V F_k^2 \, dV \right)^{1/2} \left(\int_V v_k^2 \, dV \right)^{1/2} .$$

Assuming the forces are square-integrable then, we are interested in whether the v_k are square-integrable if the strain energy of the body, given by

$$2E = \int_V \sum_{i,j,k,m=1}^{3} c^{ijkm} \varepsilon_{km}(\mathbf{u}) \varepsilon_{ij}(\mathbf{u}) \, dV ,$$

is bounded.

Since $\mathbf{v}$ is of the same class as $\mathbf{u}$, the set of all possible displacements, we study this latter set. The answer to our question is affirmative for sufficiently smooth vector functions that vanish on the boundary. This follows from inequality (6.8) written in three dimensions [see (6.11) as well, and (6.13)]. The result is one version of *Korn's inequality*: there is a constant C, which does not depend on $\mathbf{u}$ taking zero value on ∂V, such that

$$\int_V \mathbf{u} \cdot \mathbf{u} \, dV \leq C \int_V \sum_{i,j,k,m=1}^{3} c^{ijkm} \varepsilon_{km}(\mathbf{u}) \varepsilon_{ij}(\mathbf{u}) \, dV . \qquad (6.18)$$

Let us prove (6.18). By (6.13), there is a constant $m^* > 0$ such that

$$\int_V \sum_{i,j,k,m=1}^{3} c^{ijkm} \varepsilon_{km} \varepsilon_{ij} \, dV \geq m^* \int_V \Big[(\varepsilon_{11}^2 + \varepsilon_{12}^2 + \varepsilon_{22}^2)$$

$$+ (\varepsilon_{11}^2 + \varepsilon_{13}^2 + \varepsilon_{33}^2) + (\varepsilon_{22}^2 + \varepsilon_{23}^2 + \varepsilon_{33}^2) \Big] dV .$$

Now we estimate each of the three bracketed terms. Let us write

$$I_{ij} = \int_V (\varepsilon_{ii}^2 + \varepsilon_{ij}^2 + \varepsilon_{jj}^2) \, dV = \int_V [u_{i,i}^2 + \tfrac{1}{4}(u_{i,j} + u_{j,i})^2 + u_{j,j}^2] \, dV \qquad (i \neq j) .$$

For this we consider the general term

$$\int_V (u_{i,j} + u_{j,i})^2 \, dV = \int_V (u_{i,j}^2 + u_{j,i}^2 + 2u_{i,j} u_{j,i}) \, dV$$

for which, by double integration by parts, we get

$$\left| 2 \int_V u_{i,j} u_{j,i} \, dV \right| = 2 \left| \int_V u_{i,i} u_{j,j} \, dV \right| \leq \int_V (u_{i,i}^2 + u_{j,j}^2) \, dV .$$

In the last step we used the elementary inequality

$$|ab| \leq \tfrac{1}{2}(a^2 + b^2),$$

which follows from the fact that $(|a| - |b|)^2 \geq 0$. Therefore

$$2 \int_V u_{i,j} u_{j,i} \, dV \geq - \int_V (u_{i,i}^2 + u_{j,j}^2) \, dV$$

and we have

$$I_{ij} = \int_V [u_{i,i}^2 + \tfrac{1}{4}(u_{i,j}^2 + u_{j,i}^2 + 2u_{i,j} u_{j,i}) + u_{j,j}^2] \, dV$$

$$\geq \int_V [u_{i,i}^2 + \tfrac{1}{4}(u_{i,j}^2 + u_{j,i}^2 - u_{i,i}^2 - u_{j,j}^2) + u_{j,j}^2] \, dV$$

$$= \frac{3}{4} \int_V (u_{i,i}^2 + u_{j,j}^2) \, dV + \frac{1}{4} \int_V (u_{i,j}^2 + u_{j,i}^2) \, dV.$$

Uniting the terms, we get a constant M independent of $\mathbf{u}$ such that

$$\int_V \sum_{i,j,k,m=1}^{3} c^{ijkm} \varepsilon_{km}(\mathbf{u}) \varepsilon_{ij}(\mathbf{u}) \, dV \geq M \int_V \sum_{i,j=1}^{3} |u_{i,j}|^2 \, dV.$$

So by inequality (6.8) for each of the u_i we come to the needed inequality (6.18).

It can be shown that Korn's inequality also holds if only a portion of the boundary ∂V is clamped.

6.6 Theory of Elastic Plates

The relations for an elastic plate are expressed in terms of the transverse displacement w of the midplane of the plate. They are written in terms of the moments M_{ij} and the changes-of-curvature κ_{ij}, related by an equation that follows from Hooke's law and some additional assumptions on how the plate deforms:

$$M_{ij} = D^{ijkm} \kappa_{km}$$

where the D^{ijkm} are rigidity constants and the quantities

$$\kappa_{km} = -\tfrac{1}{2}(w_{,km} + w_{,mk})$$

contain second derivatives of w with respect to x_k and x_m. From the assumptions on Hooke's law constants there follow similar relations pertaining to the symmetry of the D^{ijkm},

$$D^{ijkm} = D^{jikm} = D^{kmij},$$

and positivity with some constant $c > 0$:

$$\sum_{i,j,k,m=1}^{3} D^{ijkm} \kappa_{km} \kappa_{ij} \geq c \sum_{k,m=1}^{3} \kappa_{km}^2 . \tag{6.19}$$

Rather than presenting the equilibrium equation for the plate, we will note that, similar to (6.16), for a plate occupying a domain $S \subset \mathbb{R}^2$ and having clamped edge ∂S, i.e., having

$$w\Big|_{\partial S} = 0 , \qquad \frac{\partial w}{\partial n}\Big|_{\partial S} = 0 ,$$

the weak setup turns out to be

$$\int_S \sum_{i,j,k,m=1}^{2} D^{ijkm} \kappa_{km}(w) \kappa_{ij}(v) \, dS + \int_S Fv \, dS = 0 \tag{6.20}$$

where F represents a distributed normal force.

Again,

$$\left| \int_S Fv \, dS \right| \leq \left(\int_S F^2 \, dS \right)^{1/2} \left(\int_S v^2 \, dS \right)^{1/2}$$

and we are interested in an estimate of the form

$$\int_S v^2 \, dS \leq C \int_S \sum_{i,j,k,m=1}^{2} D^{ijkm} \kappa_{km}(w) \kappa_{ij}(w) \, dS \tag{6.21}$$

with a constant C independent of v. But this follows immediately from the positivity condition (6.19) and inequality (6.8), from which there follow similar inequalities for the first derivatives:

$$\int_S v_{,i}^2 \, dS \leq C_1 \int_S \sum_{i,j,k,m=1}^{2} D^{ijkm} \kappa_{km}(w) \kappa_{ij}(w) \, dS$$

and next for the function v.

6.7 On the Weak Setup of Linear Mechanics Problems

We considered three equilibrium problems from mechanics and saw three weak setups: (6.5), (6.17), and (6.20). In linear mechanics, many problems can be reduced to the form

$$\langle u, v \rangle - F(v) = 0 , \tag{6.22}$$

where u, v can be ordinary functions or vector functions, the term $\langle u, v \rangle$ is symmetric with respect to u, v and positive so it has the properties of an inner product, and the second term $F(v)$ is linear with respect to v and hence is a linear functional. It is easy to see the correspondence between the terms of (6.22) and those of (6.5), (6.17), and (6.20). Unfortunately the spaces of smooth functions with the corresponding inner products are not complete and to study these problems we must extend the spaces. In this way we arrive at Sobolev spaces which, again, fall outside of the scope of this book. However, we should mention that in the Sobolev spaces it is easy to prove existence and uniqueness of weak solutions and to show that if the forces are square integrable, then there is an estimate

$$|F(v)| \leq c \langle v, v \rangle^{1/2} . \tag{6.23}$$

However, long before the introduction of Sobolev spaces, the weak setup was used to solve equilibrium problems numerically. So we consider this method and some results it can provide. We will present this in abstract form, but the reader can substitute the above derived expressions.

Ritz's Method. As we have said, for linear mechanics problems the weak setup follows from the minimization problem, which, in abstract form, can be stated as the problem of finding a minimizer for the functional

$$W(u) = \tfrac{1}{2}\langle u, u \rangle - F(u)$$

with a linear continuous functional F in some Hilbert space. For the membrane this functional was

$$\frac{1}{2} \int_S (u_x^2 + u_y^2) \, dS - \int_S f u \, dS ,$$

and this problem exhibits all the steps of the abstract method. First we verify that if a minimizer exists, we obtain (6.22). Assuming u is a minimizer of W, we consider $W(u + tv)$ with an arbitrary but fixed element v. Then $W(u + tv)$ becomes an ordinary function of the parameter t and takes its minimum value at $t = 0$. Hence the derivative of $W(u + tv)$ with respect to t vanishes at $t = 0$, which is precisely (6.22).

How can we use this fact? Because we cannot analytically minimize $W(u)$ for all admissible elements, in Ritz's method it is suggested that we minimize the energy functional W over a finite dimensional space in which we assume we can approximate the minimizer. In this way we arrive at a minimization problem for an ordinary function in n variables.

So we take some finite, linearly independent set $(e_1, \dots, e_n)$ of elements of the space where we are seeking the minimizer. We then seek a minimizer on the set of linear combinations

$$\sum_{k=1}^{n} c_k e_k$$

where the c_k are real scalars. For the above membrane problem, for example, $(e_1, \dots, e_n)$ is a set of functions continuously differentiable on S and vanishing on the boundary of S, which are linearly independent. Then we need to minimize

$$W\left(\sum_{k=1}^{n} c_k e_k\right) = \frac{1}{2}\left\langle \sum_{k=1}^{n} c_k e_k, \sum_{m=1}^{n} c_m e_m \right\rangle - F\left(\sum_{k=1}^{n} c_k e_k\right)$$

on the set of scalars $c_1, \ldots, c_n$. By (6.23) and the elementary inequality

$$|ca| \leq \frac{1}{4}a^2 + c^2$$

we get

$$W(u) = \frac{1}{2}\langle u, u \rangle - F(u) \geq \frac{1}{4}\langle u, u \rangle - c^2 .$$

By this,

$$\max |c_k| \to \infty \implies W\left(\sum_{k=1}^{n} c_k e_k\right) \to \infty$$

and so W is a growing function which, as a continuous function of c_k, should take its minimum value at a finite point $(c_1, \ldots, c_n)$. As it is differentiable as a quadratic polynomial we can state that at this point we get

$$\frac{\partial}{\partial c_m} W\left(\sum_{k=1}^{n} c_k e_k\right) = 0 \qquad (m = 1, \ldots, n) .$$

Explicitly, we have the system of equations

$$\begin{aligned}
c_1\langle e_1, e_1 \rangle + c_2\langle e_2, e_1 \rangle + \cdots + c_n\langle e_n, e_1 \rangle &= F(e_1) , \\
c_1\langle e_1, e_2 \rangle + c_2\langle e_2, e_2 \rangle + \cdots + c_n\langle e_n, e_2 \rangle &= F(e_2) , \\
&\vdots \\
c_1\langle e_1, e_n \rangle + c_2\langle e_2, e_n \rangle + \cdots + c_n\langle e_n, e_n \rangle &= F(e_n) .
\end{aligned} \qquad (6.24)$$

This is a linear algebraic system of equations with respect to the c_k. Its determinant, known as *Gram's determinant*, is not zero if the set $(e_1, \ldots, e_n)$ is linearly independent. Hence the system has a unique solution. We can get more information about the solution. Let $(c_1, \ldots, c_n)$ be the solution of (6.24). Multiply the mth equation by c_m and sum over m to get

$$\left\langle \sum_{k=1}^{n} c_k e_k, \sum_{m=1}^{n} c_m e_m \right\rangle = F\left(\sum_{m=1}^{n} c_m e_m\right) .$$

By (6.23) we have

$$\left\langle \sum_{k=1}^{n} c_k e_k, \sum_{m=1}^{n} c_m e_m \right\rangle \leq c \left\langle \sum_{m=1}^{n} c_m e_m, \sum_{m=1}^{n} c_m e_m \right\rangle^{1/2}$$

from which we get

$$\left\langle \sum_{m=1}^{n} c_m e_m, \sum_{m=1}^{n} c_m e_m \right\rangle^{1/2} \leq c .$$

Thus we have a uniform estimate for the approximations, independent of n. It is an a priori *estimate* for the Ritz approximations.

However, we can say something more precise about the nth Ritz approximation. By the Gram–Schmidt orthogonalization procedure, we can use $(e_1, \dots, e_n)$ to produce an orthonormal system $(g_1, \dots, g_n)$ so that

$$\langle g_i, g_j \rangle = \begin{cases} 1, & j = i, \\ 0, & j \neq i. \end{cases}$$

Clearly the minimizer of $W(\sum_{k=1}^{n} d_k g_k)$ coincides, by uniqueness, with the above sum $\sum_{k=1}^{n} c_k e_k$, but the equations of the system are much simpler:

$$\left\langle \sum_{k=1}^{n} d_k g_k, g_m \right\rangle = F(g_m),$$

and so

$$d_k = F(g_k).$$

This coefficient does not change with an increase in the order n of Ritz's system, hence we can interpret it as the kth Fourier coefficient of the solution. If we use a complete set of elements $(e_1, e_2, \dots)$ in the space, we can expect the Ritz approximations to converge to a solution of the system for any F satisfying (6.22). This can be shown if the space where we seek a solution is complete. See, e.g., [48].

6.8 Problems

6.1. For the Cauchy problem for a linear system of ordinary differential equations

$$\mathbf{x}'(t) = A(t)\mathbf{x} + \mathbf{f}(t), \qquad \mathbf{x}(a) = \mathbf{x}_0,$$

where $A(t)$ is an $n \times n$ matrix, $\mathbf{f}$ and $\mathbf{x}$ take values in $\mathbb{R}^n$, and the components of $\mathbf{f}(t), A(t)$ are piecewise continuous for $t \geq t_0$, use Gronwall's inequality to estimate the solution on $[t_0, T]$.

6.2. Using Problem 6.1, estimate on $[a, T]$ the solution to the equation

$$y^{(n)}(t) + a_1(t)y^{(n-1)}(t) + \cdots + a_n(t)y(t) = f(t)$$

with piecewise continuous coefficients and with initial conditions

$$y(a) = y_0, \quad y'(a) = y_1, \quad \dots, \quad y^{(n-1)}(a) = y_{n-1}.$$

6.3. Prove Korn's inequality in the two-dimensional case for vector functions that vanish on the boundary of their domain.

6.4. Using Ritz's method approximate a solution to the equation

$$(p(x)y'(x))' - q(x)y(x) = -f(x) \tag{6.25}$$

with boundary conditions $y(a) = 0 = y(b)$. Assume p is a given function continuously differentiable on $[a, b]$, and that q and f are continuous on $[a, b]$ with $p(x) > 0$ and $q(x) \geq 0$.

6.5. For the problem formulated in Problem 6.4, Galerkin's method is as follows. Take linearly independent functions $\phi_1, \ldots, \phi_n \in C_0^{(2)}$. Let $y_n = c_1\phi_1 + \cdots + c_n\phi_n$. Substitute this formally instead of y in (6.25), multiply by ϕ_m, and integrate with respect to x over $[a, b]$. This yields n simultaneous linear equations with respect to the variables $c_1, \ldots, c_n$; they are known as the equations of Galerkin's method (in Russia they are called the Bubnov–Galerkin equations of the nth approximation). Demonstrate that Galerkin's equations coincide with the equations of the nth approximation by Ritz's method.

Chapter 7
A Brief Introduction to Interval Analysis

7.1 Introduction

Some major advances in mathematics have occurred through the extension of existing number systems. The natural numbers were extended to the real numbers, the real numbers to the complex numbers, and so on.

In Example 1.7 we indicated that the set of closed intervals along the real line can be treated as a new class of numbers in which the real numbers are imbedded. This particular extension, when combined with the simple observation that the inequality $a \le x \le b$ is equivalent to the set membership statement $x \in [a, b]$, has opened the door to automated methods of working with inequalities.

Interval analysis, independently invented by Ramon E. Moore in the late 1950s, has developed into a multifaceted branch of mathematics with applications to global optimization, computer-assisted proofs, robotics, chemical engineering, structural engineering, computer graphics, electrical engineering, and many other areas.

In this final chapter, we offer a very modest introduction to the subject. Our presentation is informal and our only goal is to show that interval analysis represents a potent framework for working with inequalities.[1] We will not touch the subject of complex intervals. The reader is referred to Moore's books [59–61] and a few others [32, 33, 66] for more systematic expositions. Interval analysis is discussed in the context of computational functional analysis in [63].

Interval methods have found application to many areas of engineering. Space constraints prevent us from offering much of a bibliography, but in electrical engineering it is easy to find references to applications in circuit design, control and robotics, and power systems [79]. A Matlab extension called Intlab [80] is available as a convenient aid to numerical experimentation. Interval data types are also

---

[1] We like the following quote from R.D. Richtmyer, published in a review of Moore's seminal book *Interval Analysis* (Prentice Hall, 1966): "Although interval analysis is in a sense just a new language for inequalities, it is a very powerful language and is one that has direct applicability to the important problem of significance in large computations." The review appeared in *Mathematics of Computation*, Vol. 22, No. 101 (January 1968).

M.J. Cloud et al., *Inequalities: With Applications to Engineering*,
DOI 10.1007/978-3-319-05311-0_7, © Springer International Publishing AG 2014

provided by Mathematica and by certain Fortran compilers. The Interval Computations Website [88], supported by the University of Texas at El Paso, summarizes many resources available for performing interval computations.

Ramon Moore's personal account of the birth of interval analysis, originally published as an article called *The Dawning* in the journal *Reliable Computing* [62], offers an interesting look at a moment of mathematical discovery.

7.2 The Interval Number System

In interval analysis, closed intervals are commonly denoted by uppercase letters such as X. We will denote the left and right endpoints of X by $\underline{X}$ and $\overline{X}$, respectively, so that

$$X = [\underline{X}, \overline{X}] .$$

If $\underline{X} = \overline{X}$, then X is *degenerate*. Each real number a can be identified with a degenerate interval $[a, a]$, and it is in this sense that the intervals represent an extension of the real numbers.

If X and Y are intervals, and if the intersection $X \cap Y$ is not empty, then $X \cap Y$ and $X \cup Y$ are intervals given by

$$X \cap Y = [\max(\underline{X}, \underline{Y}), \min(\overline{X}, \overline{Y})] , \tag{7.1}$$

$$X \cup Y = [\min(\underline{X}, \underline{Y}), \max(\overline{X}, \overline{Y})] . \tag{7.2}$$

If $X \cap Y$ is empty, then $X \cup Y$ is not an interval. Even so, the interval on the right-hand side of (7.2) still exists; this interval contains $X \cup Y$ and is called the *interval hull* of X and Y.

The numbers

$$w(X) = \overline{X} - \underline{X}, \qquad m(X) = \tfrac{1}{2}(\underline{X} + \overline{X}),$$

are the width and midpoint, respectively, of X. If X is known to contain the exact solution of some problem, we can regard $m(X)$ as an approximation to that solution point. In this case, $w(X)$ provides error bounds for the approximation as can be seen from the midpoint-radius form

$$X = [m(X) - \tfrac{1}{2}w(X), m(X) + \tfrac{1}{2}w(X)] .$$

The absolute value of an interval X is defined as

$$|X| = \max\{|\underline{X}|, |\overline{X}|\} .$$

Example 7.1. Take $X = [1, 3]$, $Y = [2, 4]$, and $Z = [5, 9]$. The union and intersection of X and Y are

$$X \cup Y = [1, 4] , \qquad X \cap Y = [2, 3] .$$

Since $Y \cap Z$ is empty, the union $Y \cup Z$ is not an interval. However, it is contained in the interval hull of Y and Z, which is $[2, 9]$. We have $w(Y) = 2$, $m(Y) = 3$, and $|Z| = 9$. $\square$

Operations of Interval Arithmetic

A generic arithmetic operation $\oplus$ between intervals X and Y is given the following meaning:

$$X \oplus Y = \{x \oplus y : x \in X, y \in Y\} .$$

Here $\oplus$ can stand for addition, subtraction, multiplication, or division. Hence, for example, $X + Y$ is the set of all numerical sums $x + y$ where $x \in X$ and $y \in Y$. This is an interval given by

$$X + Y = [\underline{X} + \underline{Y}, \overline{X} + \overline{Y}] .$$

Similarly,

$$X - Y = [\underline{X} - \overline{Y}, \overline{X} - \underline{Y}] .$$

The product $X \cdot Y$ or XY is given by

$$XY = [\min S , \max S] \quad \text{where} \quad S = \{\underline{X}\,\underline{Y}, \underline{X}\overline{Y}, \overline{X}\,\underline{Y}, \overline{X}\,\overline{Y}\} .$$

If Y does not contain 0, then the quotient X/Y is given by

$$X/Y = X \cdot (1/Y) \quad \text{where} \quad 1/Y = [1/\overline{Y}, 1/\underline{Y}] .$$

We denote the system of closed intervals along the real line, together with these arithmetic operations, by $I(\mathbb{R})$. To illustrate, let us take $X = [1, 2]$ and $Y = [6, 8]$. Then

$$X + Y = [7, 10] , \qquad\qquad XY = [6, 16] ,$$
$$X - Y = [-7, -4] , \qquad\qquad X/Y = [\tfrac{1}{8}, \tfrac{1}{3}] .$$

The arithmetic operations in $I(\mathbb{R})$ possess many—but not all—of the properties possessed by the ordinary arithmetic operations in $\mathbb{R}$. Addition and multiplication in $I(\mathbb{R})$ are commutative and associative, but multiplication is not in general distributive over addition. Rather, the *subdistributive property*

$$X(Y + Z) \subseteq XY + XZ \tag{7.3}$$

holds (Problem 7.3). Further, the existence of additive and multiplicative inverses is not guaranteed in $I(\mathbb{R})$. By the definition of subtraction we have $X - X = 0 = [0, 0]$ only if $w(X) = 0$. Similarly, $X/X = 1$ only if $w(X) = 0$.

7.3 Outward Rounding for Rigorous Containment of Solutions

Interval analysis was designed to obtain, via machine computation, rigorous enclosures of the solutions to problems. That is, a solution to a given problem is produced
in the form of an interval guaranteed to contain the true solution. The finite nature
of machine arithmetic is addressed through *outward rounding*, i.e., rounding the left
endpoint to the closest machine number less than or equal to the exact endpoint of an
interval, and the right endpoint to the closest machine number greater than or equal
to the exact right endpoint. In outwardly rounded interval arithmetic, this procedure
is executed for every arithmetic operation—always at the last digit carried.

The machine-dependent details of implementing outward rounding are highly
technical and need not concern us here. However, some intuitive benefit may be
gained from imagining how outward rounding might be implemented by a human
using a calculator. Suppose we carry out an interval arithmetic operation on a calculator that displays nine digits and a decimal point, obtaining the interval

$$[1.23456789, 2.34567890].$$

We could, for instance, outwardly round this interval to the two-digit interval

$$[1.2, 2.4].$$

Although these intervals are not identical, any value of x contained in the first one
is certainly contained in the second one. In this way, if we are using properly implemented outwardly rounded interval arithmetic and learn that the answer y to some
particular mathematical problem lies in the interval $[3.98722, 3.98724]$, then we
know y—rigorously—to five place accuracy. The reader should consider whether
it would be preferable to compute with ordinary machine arithmetic and obtain
an answer of the form $y = 3.98722349$ but without accompanying error bounds.
A common tactic used to check for the numerical effects of finite machine representation is to switch to a higher precision arithmetic and see whether the result appears
to change. Unfortunately, it is easy to construct examples showing the unreliability
of such procedures [61].

7.4 Interval Extensions of Real-Valued Functions

Interval analysis entails computing with sets, and functions of interval arguments are
defined via set mappings. The image of an interval X under a function f is the set

$$f(X) = \{f(x) : x \in X\}. \tag{7.4}$$

We must carefully contrast this with the result of taking an *expression* for an
ordinary real function and substituting an interval for the independent variable.

The latter is called an *interval extension* of the real function. Let us denote an interval extension of a function f by F.

Example 7.2. Consider the real function f given by

$$f(x) = x^2 \qquad (-1 \le x \le 1) \,.$$

An interval extension of f is $F(X) = X \cdot X$ where X is any interval contained in $[-1, 1]$. Evaluation of $F([-1, 1])$ gives

$$F([-1, 1]) = [-1, 1] \cdot [-1, 1] = [-1, 1] \,.$$

However, $f([-1, 1])$ is obviously $[0, 1]$.   □

An interval function obtained from a real rational function f by replacing the real argument by an interval argument and the real arithmetic operations by corresponding interval operations is called a *natural interval extension* of f. It can be shown that for such an extension F,

$$f(X) \subseteq F(X) \,. \tag{7.5}$$

This is a corollary of the fundamental theorem of interval analysis, which holds more generally for any inclusion isotonic interval extension of a real function f.

We will explain the term "inclusion isotonic" in the next section. Other terms like "rational function" are defined in [61]. We merely wish to emphasize that interval analysis is used to bound solutions to problems. Whenever the sharpest (i.e., tightest) possible bound is desired, the set image $f(X)$ specified by (7.4) is the desired quantity in an interval function evaluation. It turns out that an algebraic rearrangement of a rational function expression may change the sharpness of the bound obtained.

Example 7.3. The two expressions $x(1 - x)$ and $x - x^2$ are equivalent in ordinary real arithmetic and specify the same function f for $0 \le x \le 1$. However, the corresponding natural interval extensions are not equivalent and neither one yields $f(X)$. In fact, $f(X)$ can be obtained from the expression

$$\tfrac{1}{4} - (x - \tfrac{1}{2})^2 \tag{7.6}$$

if the square is computed appropriately (i.e., not simply by the interval multiplication $(X - \tfrac{1}{2})(X - \tfrac{1}{2})$ but rather using formula (5.5) on page 38 of [61]).   □

Therefore, different real expressions for the same real function f can give rise to different interval extensions F of the function. These extensions can vary in the widths of the intervals $F(X)$ that they generate. The *dependence problem* is discussed, for example, in [33], but a basic observation is that excess width will not be generated (i.e., we will have $F(X) = f(X)$) if no variable occurs more than once in the expression used for F. This explains why the expression (7.6) (where x appears only once) is preferable to either of the expressions $x(1 - x)$ or $x - x^2$. The subdistributivity relation (7.3) makes it clear that interval expressions should be written in factored form. Unfortunately, however, the particular algebraic rearrangement of a real expression that will yield the sharpest available interval bound is not always known a priori. On a positive note, the fundamental theorem (7.5) guarantees that any valid rearrangement will yield a rigorous containment of $f(X)$.

An Application: Tolerance Analysis for Electric Circuits

Following the book [42] and the paper [79], we consider simple examples of worst-case tolerance analysis for linear, dc, resistive networks. The components making up an electric circuit are subject to variation from their nominal values because of manufacturing processes, environmental conditions, degradation with age, etc. Input variables, such as the values of voltage and current sources, are subject to measurement errors. Interval computation may be useful in placing bounds on an output quantity given bounds on multiple circuit parameters and input parameters.

Example 7.4. In Example 1.7 we used ordinary inequality manipulations to place bounds on the equivalent resistance R_e of a parallel connection of two resistors R and r (recall Fig. 1.8). Let us rework the example using interval analysis. Although Eq. (1.6) relating R_e to R and r is commonly written as

$$R_e = \frac{Rr}{R+r} , \tag{7.7}$$

the expression on the right contains both variables r and R more than once. To avoid getting an unnecessarily wide interval result because of the dependence problem, we rewrite (7.7) in the form

$$R_e = \frac{1}{1/R + 1/r} \tag{7.8}$$

before constructing the natural interval extension. For simplicity in notation, we now interpret the variables R_e, R, and r in (7.8) as intervals with

$$R = [\underline{R}, \overline{R}] \quad \text{and} \quad r = [\underline{r}, \overline{r}] .$$

Some basic interval arithmetic yields

$$R_e = [(1/\underline{R} + 1/\underline{r})^{-1} , (1/\overline{R} + 1/\overline{r})^{-1}] .$$

This interval contains all of the real values R_e given by (1.6) as the ordinary real numbers R and r are permitted to vary over their specified ranges. It is the same result we obtained in Example 1.7, where a numerical example was also given. Using the same values ($R = 1000 \pm 10\%$ and $r = 100 \pm 1\%$) and outward rounding to five places, we get a rigorous containment for R_e as $[89.189, 92.507]$.   □

Example 7.5. Suppose we want to know the node voltage V_3 in Fig. 7.1. Simple analysis using Kirchhoff's laws shows that

$$V_3 = \frac{V_1}{1 + R_1\,(1/R_2 + 1/R_3)} + \frac{V_2}{1 + R_2\,(1/R_1 + 1/R_3)}. \tag{7.9}$$

Note that we have written the expression on the right so that each variable appears just once. This is not always possible, but we should avoid the dependence problem when we can.

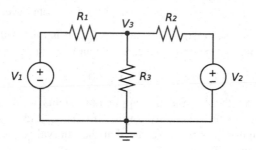

Fig. 7.1 Simple circuit for tolerance analysis

Let us assume that the voltage source values V_1 and V_2 are known precisely (see Problem 7.5 for a more general case) and that the resistances R_1, R_2, and R_3 are described using interval variables having the same names. So we now interpret the right side of (7.9) as the corresponding natural interval extension and the desired unknown V_3 as an interval to be determined. A few lines of straightforward interval arithmetic give

$$\underline{V}_3 = \frac{V_1}{1 + \overline{R}_1(1/\underline{R}_2 + 1/\underline{R}_3)} + \frac{V_2}{1 + \overline{R}_2(1/\underline{R}_1 + 1/\underline{R}_3)}$$

and

$$\overline{V}_3 = \frac{V_1}{1 + \underline{R}_1(1/\overline{R}_2 + 1/\overline{R}_3)} + \frac{V_2}{1 + \underline{R}_2(1/\overline{R}_1 + 1/\overline{R}_3)}$$

for the endpoints of V_3. As a numerical example, we could take $V_1 = 10$ and $V_2 = 5$ with $R_1 = 1000 \pm 10\,\%$, $R_2 = 100 \pm 1\,\%$, and $R_3 = 10 \pm 5\,\%$. Then $\underline{R}_1 = 900$, $\overline{R}_1 = 1{,}100$, $\underline{R}_2 = 99$, $\overline{R}_2 = 101$, $\underline{R}_3 = 9.5$, and $\overline{R}_3 = 10.5$. Outward rounding to two places, we obtain an enclosure of V_3 as $[0.50, 0.58]$. Calculations such as these are greatly facilitated by interval software such as Intlab; see [63, 80] for instructions regarding Intlab syntax, and [79] for Intlab code corresponding to Fig. 7.1. □

We have kept our examples simple, avoiding the occurrence of simultaneous systems of equations. See [42, 63] for interval methods in linear algebra.

7.5 A Few Techniques of Interval Analysis

Fixed Point Iteration with Intervals

In Example 4.7 we discussed a scheme for calculating the square root of a positive number. The method was fixed point iteration in ordinary arithmetic. Let us show how this problem can be approached using interval arithmetic.

Definition 7.1. Let F be an interval extension of a function f, and let X, Y be intervals. We say that F is *inclusion isotonic* if $F(X) \subseteq F(Y)$ whenever $X \subseteq Y$.

We mentioned inclusion isotonicity in Sect. 7.4; it is one of the keys to understanding interval analysis. Suppose we carry out an interval iterative process

$$X_{n+1} = F(X_n) \qquad (n = 0, 1, 2, \ldots) \tag{7.10}$$

where it happens that the first function application yields an interval containment: $X_1 \subseteq X_0$. Then inclusion isotonicity implies $F(X_1) \subseteq F(X_0)$ or in other words $X_2 \subseteq X_1$. Continuing in this way, we see that the interval sequence generated by (7.10) will be nested with

$$X_0 \supseteq X_1 \supseteq X_2 \supseteq X_3 \supseteq \cdots . \tag{7.11}$$

Now suppose we start with an ordinary iterative scheme

$$x_{n+1} = f(x_n) \qquad (n = 0, 1, 2, \ldots)$$

where f has a fixed point x^*. We replace this scheme by the interval scheme (7.10), taking F as an inclusion isotonic extension of f and choosing a starting interval X_0 that contains x^*. Since $x^* \in X_0$, we will have

$$x^* = f(x^*) \in f(X_0) \subseteq F(X_0) = X_1$$

and therefore x^* will also lie in X_1, the first interval produced by (7.10). Repetition of this reasoning shows that x^* will lie in X_k for all $k = 0, 1, 2, \ldots$. In other words, x^* will lie in the intersection $\cap_k X_k$ of all the intervals generated. Of course, a nested sequence of intervals $\{X_k\}$ may not have $w(X_k) \to 0$ as $k \to \infty$; a constant sequence of intervals would also satisfy (7.11). But in certain cases it is possible to show that $w(X_k) \to 0$ and hence the interval sequence $\{X_k\}$ converges to the degenerate interval $[x^*, x^*]$.

Example 7.6 ([58, 64]). The equation $x^2 = 2$ in ordinary arithmetic is equivalent to the equation

$$x = 1 + \frac{1}{1 + x}$$

and the function $f(x)$ on the right-hand side has fixed point $\sqrt{2}$. Replacing x by X, we produce an interval extension of f given by

$$F(X) = 1 + \frac{1}{1 + X} .$$

When started with $X_0 = [1, 2]$, the iteration (7.10) produces

$$X_1 = [\tfrac{4}{3}, \tfrac{3}{2}], \quad X_2 = [\tfrac{7}{5}, \tfrac{10}{7}], \quad X_3 = [\tfrac{24}{17}, \tfrac{17}{12}], \quad X_4 = [\tfrac{41}{29}, \tfrac{58}{41}], \quad \cdots$$

(Problem 7.4). To show that we are actually closing in on $\sqrt{2}$, we examine the width of $F(X)$ given by

$$w(F(X)) = w\left(1 + \frac{1}{1 + X}\right).$$

But it is easily verified (Problem 7.2) that if X, Y are intervals and α, β are positive real numbers, then

$$w(\alpha X + \beta Y) = \alpha w(X) + \beta w(Y)$$

so

$$w(F(X)) = w(1) + w\left(\frac{1}{1 + X}\right) = w\left(\frac{1}{1 + X}\right). \tag{7.12}$$

Furthermore (Problem 7.2), for any interval I such that $0 \notin I$,

$$w(1/I) \le |1/I|^2 w(I)$$

where $|[a, b]| = \max\{|a|, |b|\}$. Note that if $0 \notin [a, b]$, then

$$|1/[a, b]| = |[1/b, 1/a]| = \max\{|1/b|, |1/a|\} = \begin{cases} 1/a, & 0 < a \le b, \\ -1/b, & a \le b < 0. \end{cases}$$

Returning to (7.12), we have

$$w(F(X)) \le \left|\frac{1}{1 + X}\right|^2 w(1 + X)$$

where $w(1 + X) = w(X)$ and, assuming $X \subseteq [1, 2]$,

$$\left|\frac{1}{1 + X}\right| = \left|\frac{1}{1 + [\underline{X}, \overline{X}]}\right| = \left|\frac{1}{[1 + \underline{X}, 1 + \overline{X}]}\right| = \left|\left[\frac{1}{1 + \overline{X}}, \frac{1}{1 + \underline{X}}\right]\right| = \frac{1}{1 + \underline{X}} \le \frac{1}{2}.$$

Therefore

$$w(F(X)) \le \tfrac{1}{4} w(X).$$

Applying this to the sequence $\{X_k\}$, we get

$$w(X_k) = w(F(X_{k-1})) \le (\tfrac{1}{4})^1 w(X_{k-1}) = (\tfrac{1}{4})^1 w(F(X_{k-2}))$$
$$\le (\tfrac{1}{4})^2 w(X_{k-2}) = \cdots \le (\tfrac{1}{4})^k w(X_0) \to 0 \text{ as } k \to \infty.$$

We are assured that $\{X_k\}$ converges to the degenerate interval containing $\sqrt{2}$. $\quad\square$

The iteration scheme (7.10) can be improved by intersecting each $F(X_k)$ with the previously obtained interval X_k:

$$X_{k+1} = F(X_k) \cap X_k \qquad (k = 0, 1, 2, \ldots) .$$

This form provides information even if X_1 is not contained in X_0. For example, if the intersection is empty, then there is no fixed point in X_0. See, e.g., [61] for further information on interval fixed point methods.

Interval Newton Method

The reader is assumed to be familiar with Newton's method and its initial guess sensitivity. An interval Newton method is available for solving nonlinear equations. It can be used in an algorithm guaranteed to find *all* roots in some initial interval. The interval Newton method also exhibits quadratic convergence.

Let us formulate the iteration rule for one equation in one unknown. We seek a solution of the equation

$$f(x) = 0 \tag{7.13}$$

in an interval X, assuming f is continuously differentiable. The iteration rule is

$$X_{k+1} = X_k \cap \left\{ m(X_k) - \frac{f(m(X_k))}{F'(X_k)} \right\} \qquad (k = 0, 1, 2, \ldots) . \tag{7.14}$$

The idea behind (7.14) is as follows. If $x, y_0 \in X$ and x satisfies (7.13), then by Corollary 2.4 we can write

$$x = y_0 - \frac{f(y_0)}{f'(\xi)}$$

where ξ lies between x and y_0. Now let $F'(X)$ be an inclusion isotonic interval extension of $f'(x)$. Since $\xi \in X$, we have $f'(\xi) \in F'(X)$ and therefore

$$x \in y_0 - \frac{f(y_0)}{F'(X)} .$$

As indicated in (7.14), we ordinarily take y_0 to be the midpoint $m(X)$ of X. The intersection with X_k is included, among other reasons, to speed convergence of the algorithm [61].

Example 7.7. We can find $\sqrt{2}$ by defining $f(x) = x^2 - 2$ and using $F'(X) = 2X$. In this case the iteration rule (7.14) looks like

$$X_{k+1} = X_k \cap \left\{ m(X_k) - \frac{[m(X_k)]^2 - 2}{2X_k} \right\} \qquad (k = 0, 1, 2, \ldots) .$$

Starting with $X_0 = [1, 2]$, we obtain

$$X_1 = [\tfrac{11}{8}, \tfrac{23}{16}] = [1.375, 1.4375],$$

$$X_2 = [\tfrac{181}{128}, \tfrac{3983}{2816}] = [1.41406\ldots, 1.41441\ldots],$$

$$\vdots$$

which quadratically close in on $\sqrt{2}$. To illustrate how to find *all* zeros in a given interval, consider $X_0 = [-2, 2]$. We have $m(X_0) = 0$, and so

$$X_1 = [-2, 2] \cap \{1/[-2, 2]\}$$

$$= [-2, 2] \cap \left\{(-\infty, -\tfrac{1}{2}] \cup [\tfrac{1}{2}, \infty]\right\}$$

$$= [-2, -\tfrac{1}{2}] \cup [\tfrac{1}{2}, 2]$$

and we continue by processing the two separate subintervals $[-2, -\tfrac{1}{2}]$ and $[\tfrac{1}{2}, 2]$. We will find both of the roots in the starting interval, namely both the negative and the positive square roots of 2. The interval Newton method can also be applied to systems of equations in n-dimensions.   □

Example 7.8. Consider the circuit of Fig. 7.2.

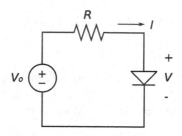

Fig. 7.2 Simple nonlinear circuit for application of the interval Newton method

Assuming the diode is described for positive voltage V by the current–voltage relationship

$$I = I_s \exp(V/nV_T)$$

where I_s, n, and V_T are constants, we can apply Kirchhoff's voltage law and obtain the equation

$$I_s \exp(V/nV_T) - \frac{1}{R}(V_0 - V) = 0$$

where V is the unknown. We can define the left-hand side as a function $f(V)$ and seek a zero in a given starting interval. With $I_s = 1 \times 10^{-14}$, $n = 1.8$, $V_T = 26 \times 10^{-3}$, $R = 400{,}000$, and a starting interval $[0.8, 0.9]$, we find that $V \in [0.823, 0.824]$.   □

Refinement

Let us repeat the fundamental theorem (7.5), which guarantees that if F is an inclusion isotonic interval extension of a real function f, then

$$f(X) \subseteq F(X) .$$

The set $f(X)$ is the image of X under the function f: i.e., the set of all values $f(x)$ for $x \in X$. In Example 7.2 we saw that under certain circumstances $F(X)$ can provide rather loose bounds for $f(X)$. However, the bounds available from an interval extension may be tightened using *refinement*, a process in which we partition the interval X into subintervals $X_1, \ldots, X_n$ and take the interval hull of $F(X_1), \ldots, F(X_n)$.

To see why refinement might produce a tighter enclosure of $f(X)$, let us consider a simple argument [42]. Write

$$X = \cup_{i=1}^{n} X_i$$

and operate on both sides with f:

$$f(X) = f(\cup_{i=1}^{n} X_i) = \cup_{i=1}^{n} f(X_i) .$$

Since $f(X_i) \subseteq F(X_i)$ for each i, we have

$$f(X) \subseteq \cup_{i=1}^{n} F(X_i) . \qquad (7.15)$$

On the other hand, we have $X_i \subseteq X$ for each i, so by inclusion isotonicity $F(X_i) \subseteq F(X)$ and therefore

$$\cup_{i=1}^{n} F(X_i) \subseteq F(X) . \qquad (7.16)$$

Relations (7.15) and (7.16) show that the union of the $F(X_i)$ always contains $f(X)$ and is never wider than $F(X)$. In fact a stronger statement can be made, but first we illustrate refinement with an example from Moore's dissertation [64].

Example 7.9. Let $f(x) = x(1 - x)$ for $0 \le x \le 1$ and take $F(X) = X(1 - X)$ where $X \subseteq [0, 1]$. In this case $f([0, 1]) = [0, \frac{1}{4}]$, but $F([0, 1]) = [0, 1]$. To implement refinement, we can write

$$[0, 1] = \cup_{i=1}^{n} [\tfrac{i-1}{n}, \tfrac{i}{n}]$$

and be assured, by (7.15), that

$$f([0, 1]) \subseteq \cup_{i=1}^{n} F\left([\tfrac{i-1}{n}, \tfrac{i}{n}]\right) .$$

Moore showed that

$$f([0, 1]) \subseteq [0, \tfrac{1}{4}] + \begin{cases} [0, \tfrac{1}{2n}], & n \text{ even} , \\ [0, \tfrac{1}{2n} + \tfrac{1}{4n^2}], & n \text{ odd} . \end{cases}$$

Observe that the refinements close down on $[0, \frac{1}{4}]$ with increasing n. In this example we can produce *arbitrarily* tight enclosures using refinement.   □

Example 7.9 illustrates not only relations (7.15) and (7.16), but the following theorem. Define the *excess width* of an interval extension $F(X)$ of a real function $f(x)$ as the number $w(F(X)) - w(f(X))$.

Theorem 7.1. *Suppose $F(X)$ is an inclusion isotonic, Lipschitz, interval extension of $f(x)$. If $F_{(n)}(X)$ is a refinement of $F(X)$ produced from a uniform subdivision of X into n equal subintervals, then the excess width of $F_{(n)}(X)$ is $O(1/n)$.*

See [61] for a proof, the definition of the term "Lipschitz interval extension," refinement code for the Intlab extension to Matlab, and additional examples.

Polynomial Enclosure of the Solution to an Operator Equation

The solutions to differential and integral equations can be approximated by enclosing them in interval polynomials. We illustrate with a simple example from [32].

Example 7.10. Consider the initial value problem

$$y'(x) = y^2(x) , \qquad y(0) = 1 . \tag{7.17}$$

This can be rewritten as the integral equation

$$y(x) = 1 + \int_0^x y^2(\xi)\, d\xi . \tag{7.18}$$

With (7.18) in mind, we try the sequence of interval polynomials $\{P_k(x)\}$ given by

$$P_{k+1}(x) = 1 + \int_0^x P_k^2(\xi)\, d\xi \qquad (k = 0, 1, 2, \ldots)$$

where $P_0(x) = [1, d]$ is a constant interval polynomial. The condition

$$P_1(x) \subseteq P_0(x) \text{ for } x \in [0, a] \text{ and some } a > 0 \tag{7.19}$$

is sufficient to guarantee that $P_{k+1}(x) \subseteq P_k(x)$ for every k and all $x \in [0, a]$.
    But

$$P_1(x) = 1 + \int_0^x [1, d]^2 \, d\xi = 1 + [1, d^2]x = [1 + x^2, 1 + d^2 x] ,$$

hence to implement (7.19) we seek a pair d, a such that $1 \le 1 + x^2$ and $1 + d^2 x \le d$ for all $x \in [0, a]$. The first of these inequalities is satisfied automatically; the second one requires $1 + d^2 a \le d$. Continuing to follow [32], we take $d = 2$ and $a = 1/4$.

Finally, starting with the constant interval polynomial

$$P_0(x) = [1, d] = [1, 2] \,,$$

we generate

$$P_1(x) = 1 + \int_0^x P_0^2(\xi)\, d\xi = 1 + \int_0^x [1, 2]^2(\xi)\, d\xi = 1 + [1, 4]x \,,$$

and

$$P_2(x) = 1 + \int_0^x P_1^2(\xi)\, d\xi$$

$$= 1 + \int_0^x (1 + [1, 4]\xi)^2\, d\xi$$

$$= 1 + \int_0^x (1 + 2[1, 4]\xi + [1, 4]^2\xi^2)\, d\xi$$

$$= 1 + \int_0^x (1 + 2[1, 4]\xi + [1, 16]\xi^2)\, d\xi$$

$$= 1 + x + [1, 4]x^2 + [\tfrac{1}{3}, \tfrac{16}{3}]x^3 \,,$$

and so on. It can be shown that this sequence $\{P_k(x)\}$ given by

$$P_0(x) = [1, 2]$$
$$P_1(x) = 1 + [1, 4]x$$
$$P_2(x) = 1 + x + [1, 4]x^2 + [\tfrac{1}{3}, \tfrac{16}{3}]x^3$$

$$\vdots$$

converges uniformly to the solution $y(x)$ of the original problem (7.17) on the interval $[0, \tfrac{1}{4}]$ and that each $P_k(x)$ represents an *enclosure* for $y(x)$ in the sense that $y(x) \in P_k(x)$ on $[0, \tfrac{1}{4}]$. See Problem 7.6.   □

Note that we have integrated an interval function in this example. See [60, 61] for further background on interval integrals and the solution of operator equations by interval methods.

7.6 Further Reading

Fundamentally, interval analysis is about computing with sets and producing rigorous containments of solutions via machine computation. Interval methods have been developed for global optimization, the solution of integral equations, matrix problems, the solution of initial and boundary value problems for systems of differential equations, and so on. The Interval Computations Website [88] and the more recent book [61] can supply the reader with further information and more specific routes to learning about contemporary applications.

7.7 Problems

7.1. Let A, B, C, D be intervals with $\underline{A}, \underline{B}, \underline{C}, \underline{D} > 0$. Find endpoint expressions for the intervals

(a) $1 + B/A$,

(b) $AB/(B + C)$,

(c) $AB/(CD)$,

(d) $A + B + C$,

(e) $(B + C)/A$,

(f) $(A + BC)/D$,

(g) $C(A + 1/B)$,

(h) $A/B + C/D$,

(i) $1/[1 + A(1/B + 1/C)]$.

7.2. Prove the following statements about interval width and absolute value:

(a) $w(XY) \le |X| w(Y) + |Y| w(X)$,

(b) $|\alpha X| = |\alpha| |X|$,

(c) $w(\alpha X) = |\alpha| w(X)$,

(d) $|X + Y| \le |X| + |Y|$,

(e) $w(1/X) \le |1/X|^2 w(X)$,

(f) $|XY| = |X| |Y|$,

(g) If $X \subseteq Y$, then $w(X) \le w(Y)$,

(h) $w(X + Y) = w(X) + w(Y)$,

(i) $w(X - Y) = w(X) + w(Y)$,

(j) $w(XY) \ge |X| w(Y)$,

(k) $w(XY) \ge \max\{|X| w(Y), |Y| w(X)\}$.

7.3. Prove (7.3).

7.4. Verify the interval arithmetic computations in Example 7.6.

7.5. Rework Example 7.5 taking the voltage sources V_1 and V_2 as interval variables.

7.6. Verify that the solution to the initial value problem of Example 7.10 is

$$y(x) = \frac{1}{1 - x}.$$

Plot this function on $[0, \frac{1}{4}]$ along with the bounding functions specified by $P_0(x)$, $P_1(x)$, and $P_2(x)$.

Appendix A
Hints for Selected Problems

In this appendix we use "iff" as an abbreviation for "if and only if." Equations are occasionally tagged with asterisks for reference [e.g., as (*) or (**)]. All such equation labels are purely local and have meaning only within the hint for a given problem.

1.1 Adding the inequalities $x \leq |x|$ and $y \leq |y|$, we get

$$x + y \leq |x| + |y| . \qquad (*)$$

Replacing x by $-x$ and y by $-y$ in (*), we get

$$x + y \geq -(|x| + |y|) . \qquad (**)$$

Combining (*) and (**), we get (1.12). The stated conditions for equality are easily checked.

1.2

(a) Equivalent to $(x - y)^2 \geq 0$.

(b) Equivalent to $(wz - xy)^2 \geq 0$.

(c) Multiply the results

$$x^2 + y^2 \geq 2xy , \quad y^2 + z^2 \geq 2yz , \quad z^2 + x^2 \geq 2zx ,$$

by z^2, x^2, and y^2, respectively, and add.

(d) Equivalent to $(x - y)^2 \geq 0$.

(e) Use a double application of (1.34):

$$x^4 + y^4 + z^4 = (x^2)^2 + (y^2)^2 + (z^2)^2 \geq x^2 y^2 + y^2 z^2 + z^2 x^2 = (xy)^2 + (yz)^2 + (zx)^2$$
$$\geq (xy)(yz) + (yz)(zx) + (zx)(xy) = xyz(x + y + z) .$$

(f) Equivalent to $(x - yz)^2 \geq 0$.

(g) Equivalent to

$$\left(\frac{x}{\sqrt{2}} - \frac{y}{\sqrt{2}} \right)^2 + \left(\frac{y}{\sqrt{2}} - \frac{z}{\sqrt{2}} \right)^2 + \left(\frac{z}{\sqrt{2}} - \frac{x}{\sqrt{2}} \right)^2 \geq 0 .$$

Equality holds iff $x = y = z$.

1.3 Start from $(x - 1)^2 \geq 0$, noting that x must be positive in order to obtain $x + 1/x \geq 2$ from this.

M.J. Cloud et al., *Inequalities: With Applications to Engineering*,
DOI 10.1007/978-3-319-05311-0, © Springer International Publishing AG 2014

1.4 Start with the inequality

$$\left(\sqrt{\varepsilon}a - \frac{b}{\sqrt{\varepsilon}}\right)^2 \geq 0 .$$

Expanding the square, we get

$$ab \leq \frac{\varepsilon a^2}{2} + \frac{b^2}{2\varepsilon} .$$

Changing b to $-b$, we get

$$-ab \leq \frac{\varepsilon a^2}{2} + \frac{b^2}{2\varepsilon} ,$$

which completes the proof. The result appears in various forms. For example, we can replace ε by 2ε and write it as

$$|ab| \leq \varepsilon a^2 + \frac{b^2}{4\varepsilon} .$$

This form, with $\varepsilon = 1$, will be used on p. 176.

1.5 It is helpful to remember that if $u < v$, then for any $n \in \mathbb{N}$ we have

$$u^{2n+1} < v^{2n+1} \quad \text{and} \quad {}^{2n+1}\!\sqrt{u} < {}^{2n+1}\!\sqrt{v} .$$

If $0 \leq u < v$, we also have

$$u^n < v^n \quad \text{and} \quad \sqrt[n]{u} < \sqrt[n]{v} .$$

(a)   Equivalent to

$$f(x) > g(x)^{2n+1} .$$

(b)   Equivalent to

$$f(x) < g(x)^{2n+1} .$$

(c)   Equivalent to

$$g(x) \geq 0 \text{ and } f(x) > g(x)^{2n} \quad or \quad g(x) < 0 \text{ and } f(x) \geq 0 .$$

(d)   Equivalent to

$$g(x) > 0 \quad \text{and} \quad 0 \leq f(x) < g(x)^{2n} .$$

(e)   Equivalent to

$$g(x) > 0 \quad \text{and} \quad f(x) > g(x)^{2n} .$$

(f)   Equivalent to

$$g(x) > 0 \text{ and } 0 \leq f(x) < g(x)^{2n} \quad or \quad g(x) < 0 \text{ and } f(x) \geq 0 .$$

1.6 Let us review some facts about the logarithmic and exponential functions. The function $\log_b x$ is continuous for $x > 0$. It is increasing if $b > 1$ and decreasing if $0 < b < 1$. As an example, we sketch the second of these cases in Fig. A.1.

The exponential function a^x is defined only for $a > 0$. If $a > 1$, then a^x is increasing for all x; furthermore, we have $0 < a^x < 1$ for $x < 0$, and $a^x > 1$ for $x > 0$. If $0 < a < 1$, then a^x is decreasing for all x; furthermore, we have $a^x > 1$ for $x < 0$, and $0 < a^x < 1$ for $x > 0$.

(a)   *The inequality $a^x > b$. If $b > 0$ and $a > 1$, the solution is $x > \log_a b$. If $b > 0$ and $0 < a < 1$, the solution is $x < \log_a b$. If $b \leq 0$, the inequality holds for all $x \in \mathbb{R}$.*
        The inequality $a^x < b$. If $b > 0$ and $a > 1$, the solution is $x < \log_a b$. If $b > 0$ and $0 < a < 1$, the solution is $x > \log_a b$. If $b \leq 0$, the inequality never holds.

(b)   If $a > 1$, the solution is $x > \alpha$. If $0 < a < 1$, the solution is $x < \alpha$.

(c)   If $a > 1$, the inequality is equivalent to $f(x) > g(x)$. If $0 < a < 1$, it is equivalent to $f(x) < g(x)$.

(d)   The inequality holds iff $f(x) > 1$ and $h(x) < g(x)$ *or* $0 < f(x) < 1$ and $h(x) > g(x)$.

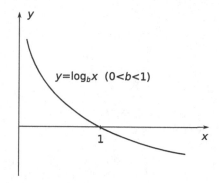

$y=\log_b x \quad (0<b<1)$

Fig. A.1 Logarithmic function with base less than unity

(e) The inequality holds iff $h(x) > 0$ and $0 < f(x) < g(x)$ *or* $h(x) < 0$ and $0 < g(x) < f(x)$.

(f) *The inequality* $\log_a x > b$. *If* $a > 1$, *the solution is* $x > a^b$. *If* $0 < a < 1$, *it is* $0 < x < a^b$.
The inequality $\log_a x < b$. *If* $a > 1$, *the solution is* $0 < x < a^b$. *If* $0 < a < 1$, *it is* $x > a^b$.

(g) *The inequality* $\log_a x > \log_a \alpha$. *If* $a > 1$, *the solution is* $x > \alpha$. *If* $0 < a < 1$, *it is* $0 < x < \alpha$.
The inequality $\log_a x < \log_a \alpha$. *If* $a > 1$, *the solution is* $0 < x < \alpha$. *If* $0 < a < 1$, *it is* $x > \alpha$.

(h) If $a > 1$, the inequality is equivalent to $0 < f(x) < g(x)$. If $0 < a < 1$, it is equivalent to
$0 < g(x) < f(x)$.

(i) *The inequality* $\log_{g(x)} f(x) > 0$ *holds iff* $g(x) > 1$ *and* $f(x) > 1$ *or* $0 < g(x) < 1$ *and*
$0 < f(x) < 1$.
The inequality $\log_{g(x)} f(x) < 0$ *holds iff* $g(x) > 1$ *and* $0 < f(x) < 1$ *or* $0 < g(x) < 1$ *and*
$f(x) > 1$.

(j) The inequality holds iff $h(x) > 1$ and $0 < f(x) < g(x)$ *or* $0 < h(x) < 1$ and $0 < g(x) < f(x)$.

1.7

(a) Use the method of intervals as in Example 1.14. Factor $x^2 - 2x - 3$ as $(x + 1)(x - 3)$. For
$x > 3$, the product is clearly positive. In the interval $-1 < x < 3$, the product is negative;
the factor $x - 3$ changes sign as we pass through $x = 3$, while $x + 1$ does not. As we pass
through $x = -1$, the product changes sign again. The solution set is $(-1, 3)$.

(b) Use the method of intervals; $(x + 4)^2$ never changes sign, so it can be ignored.

(c) $(-1, 1)$, by the method of intervals. *Remark.* We have noted that the inequalities $f(x)/g(x) >$
0 and $f(x)g(x) > 0$ have the same solution set. We should caution, however, that this is not
true for the weak versions $f(x)/g(x) \geq 0$ and $f(x)g(x) \geq 0$. The second inequality of this
pair may be treated as a preliminary step, but it is clear that any zeros of $g(x)$ must be
excluded as they will fail to satisfy the first inequality of the pair. Analogous remarks hold
for inequalities of the opposite sense ($<$ and $\leq$).

(d) $(-\infty, 0)$.

(e) $(-1, \infty)$.

(f) Rearrange as $x^2 - 1 \geq |x|$ and graph the functions on the left- and right-hand sides. The
solution set is $(-\infty, (-1 - \sqrt{5})/2] \cup [(1 + \sqrt{5})/2, \infty)$.

(g) Equivalent to the pair of conditions $x > 0$ and $-x < x^2 - x < x$. The solution set is $(0, 2)$.

(h) $(-\infty, -1) \cup (0, 1) \cup (1, \infty)$.

(i) $(0, 1) \cup (1, \infty)$.

(j) $[-2, 2)$.

(k) $(-1/2, \infty)$.

(l) $[-1, (\sqrt{5} - 1)/2)$.

(m)   Write as $(1/2)^{x+1} \geq (1/2)^0$. The solution set is $(-\infty, -1]$.

(n) $[-2, -1/2] \cup [1/2, 2]$.

(o) $(0, 1)$.

(p) $(-\sqrt{-a}, 0)$ if $a < 0$; $(0, \sqrt{a})$ if $a > 0$; $\emptyset$ if $a = 0$.

(q) $1/a < x < +\infty$ if $a > 0$; $-\infty < x < 1/a$ if $a < 0$; $\emptyset$ if $a = 0$.

(r) $(-\infty, 0)$ if $a \leq 0$; $(-1/\sqrt{a}, 0) \cup (1/\sqrt{a}, \infty)$ if $a > 0$.

(s) $\mathbb{R}$ if $a \geq 0$; $[-\sqrt{-1/a}, \sqrt{-1/a}]$ if $a < 0$.

(t) $(-\infty, a/2]$ if $a < 0$; $\mathbb{R}$ if $a \geq 0$.

(u) $\mathbb{R}$ if $a < 0$; $(-\infty, -1) \cup (-1, 1) \cup (1, \infty)$ if $a = 0$; $(-\sqrt{1-a}, \sqrt{1-a}) \cup (-\infty, -\sqrt{1+a}) \cup$
        $(\sqrt{1+a}, \infty)$ if $0 < a < 1$; $(-\infty, -\sqrt{1+a}) \cup (\sqrt{1+a}, \infty)$ if $a \geq 1$.

(v) $[-1, \infty)$ if $a \in (-\infty, 0]$; $[-1, \frac{1}{a^2} - 1)$ if $a \in (0, \infty)$.

(w) $\mathbb{R}$ if $a \leq -1$; $(-\infty, \frac{a-1}{2}]$ if $a > -1$.

(x) $\mathbb{R}$ if $-1 \leq a \leq 1$; $(\frac{-1}{|a|-1}, \frac{1}{|a|-1})$ otherwise.

(y) $\emptyset$ if $a \leq 0$; $(-\sqrt{1+a}, -\sqrt{1-a}) \cup (\sqrt{1-a}, \sqrt{1+a})$ if $0 < a < 1$; $(-\sqrt{1+a}, 0) \cup (0, \sqrt{1+a})$
        if $a = 1$; $(-\sqrt{1+a}, \sqrt{1+a})$ if $a > 1$.

(z) $(0, 2^{a+2})$ for any a.

Additional remarks about inequalities with absolute values. (1) In the common domain of definition of the functions $f(x)$ and $g(x)$, the inequality

$$|f(x)| + |g(x)| > 0$$

can *fail* to hold only at points x such that both $f(x) = 0$ and $g(x) = 0$. For example, the inequality

$$|\ln x - 1| + |x - e| > 0$$

holds for all positive x except those that satisfy the system

$$\ln x - 1 = 0 \quad \text{and} \quad x - e = 0 \,.$$

So the solution set is $(0, e) \cup (e, \infty)$. (2) If $f(x)$, $g(x)$, and $h(x)$ are polynomials, then an inequality of the form

$$|f(x)| + |g(x)| > h(x)$$

can be broken into cases considered on intervals. Take, for instance,

$$|x| + |x - 1| > 2 \,.$$

Setting $x = 0$ and $x - 1 = 0$, we are led to consider the inequality on the intervals $(-\infty, 0)$, $(0, 1)$, and $(1, \infty)$. On the first of these we have $|x| = -x$ and $|x - 1| = 1 - x$, so the given inequality is equivalent to $-x + 1 - x > 2$. This gives $x < -1/2$. Hence part of the solution set is $(-\infty, -1/2)$. For $0 < x < 1$ we have $|x| = x$ but $|x - 1| = 1 - x$ and the given inequality is equivalent to $x + 1 - x > 2$. This has no solution. For $x > 1$ we have $|x| = x$ and $|x - 1| = x - 1$ and the given inequality is equivalent to $x + x - 1 > 2$. This gives $x > 3/2$. So the rest of the solution set is $(3/2, \infty)$.

1.8

(a) $\mathcal{P}(1)$ holds as $2^3 > 2(1) + 5$. Assuming $\mathcal{P}(n)$ holds, we have $2 \cdot 2^{n+2} > 2(2n + 5)$ so that

$$2^{(n+1)+2} > 2(n + 1) + 5 + \{2(2n + 5) - [2(n + 1) + 5]\} = 2(n + 1) + 5 + (2n + 3) > 2(n + 1) + 5 \,.$$

This is $\mathcal{P}(n + 1)$.

(b) $\mathcal{P}(3)$ holds as $2^3 > 2(3) + 1$. Assuming $\mathcal{P}(n)$ holds, we have $2 \cdot 2^n > 2(2n + 1)$ so that

$$2^{n+1} > 2(n + 1) + 1 + \{2(2n + 1) - [2(n + 1) + 1]\} = 2(n + 1) + 1 + (2n - 1) > 2(n + 1) + 1 .$$

This is $\mathcal{P}(n + 1)$.

(c)   Verify the relation

$$\frac{1}{\sqrt{n + 1}} < 2\sqrt{n + 1} - 2\sqrt{n}$$

and add it to the induction hypothesis $\mathcal{P}(n)$.

(e) $\mathcal{P}(2)$ holds as $2!\,4! > [3!]^2$. To show that $\mathcal{P}(n) \implies \mathcal{P}(n + 1)$, we start with the obvious relation

$$(n + 2)(n + 3) \cdots (2n + 2) > (n + 2)^{n+1}$$

where there are $(2n + 2) - (n + 1) = n + 1$ factors on the left-hand side. Multiply both sides by $(n + 1)!$ to get

$$(2n + 2)! > (n + 2)^{n+1}(n + 1)!$$

and then multiply this inequality by the induction hypothesis to get

$$2!\,4! \cdots (2n)!(2n + 2)! > [(n + 1)!]^n(n + 2)^{n+1}(n + 1)!$$

The left-hand side is $2!\,4! \cdots [2(n + 1)]!$ while the right-hand side is $\{[(n + 1) + 1]!\}^{n+1}$.

(f) $\mathcal{P}(2)$ holds. To show that $\mathcal{P}(n) \implies \mathcal{P}(n + 1)$, we multiply the induction hypothesis by the easily verified relation

$$4(n + 1)^2 < \frac{(n + 2)(2n + 2)(2n + 1)}{n + 1}$$

to get

$$4^{n+1}[(n + 1)!]^2 < [(n + 1) + 1][2(n + 1)]!$$

which is $\mathcal{P}(n + 1)$.

(g) $\mathcal{P}(2)$ is

$$\frac{a_2^2}{a_1} \geq 4(a_2 - a_1)$$

which is equivalent to $(a_2 - 2a_1)^2 \geq 0$. To show that $\mathcal{P}(k) \implies \mathcal{P}(k + 1)$, start with

$$\frac{a_{n+1}^2}{a_n} \geq 4(a_{n+1} - a_n)$$

(equivalent to $(a_{n+1} - 2a_n)^2 \geq 0$) and add this inequality to the induction hypothesis to get

$$\frac{a_2^2}{a_1} + \frac{a_3^2}{a_2} + \cdots + \frac{a_n^2}{a_{n-1}} + \frac{a_{n+1}^2}{a_n} \geq 4(a_{n+1} - a_n) + 4(a_n - a_1)$$

Since the right-hand side is $4(a_{n+1} - a_1)$, this is $\mathcal{P}(n + 1)$.

(i)   Because of the constant on the right-hand side, we prove the stronger inequality

$$\frac{1}{2^2} + \frac{1}{3^2} + \cdots + \frac{1}{n^2} < 1 - \frac{1}{n} \qquad (n \geq 2) .$$

This is straightforward by induction.

(j)   The inequality holds for $k = 3$. Assume it holds for $k = n$:

$$2^{\frac{1}{2}n(n-1)} > n!$$

Multiply through by 2^n:

$$2^n \cdot 2^{\frac{1}{2}n(n-1)} > 2^n \cdot n!$$

The exponent on the left-hand side is $n + \frac{1}{2}n(n-1) = \frac{1}{2}(n+1)n$, so by (1.35) we get

$$2^{\frac{1}{2}(n+1)n} > 2^n \cdot n! > (n+1)n! = (n+1)!$$

(k)   Put $n = k$ into the inequality and create a proposition $\mathcal{P}(k)$. Proposition $\mathcal{P}(1)$ is $1 + x \le 1 + x$ and thus holds trivially. It remains to show that $\mathcal{P}(k) \implies \mathcal{P}(k+1)$. Hence we assume $\mathcal{P}(k)$ holds, and multiply both sides by the positive quantity $1 + x$:

$$(1+x)^{k+1} \le (1+x) + (1+x)(2^k - 1)x \le (1+x) + 2(2^k - 1)x = 1 + (2^{k+1} - 1)x .$$

This is $\mathcal{P}(k+1)$, as required.

(l)   Induction isn't needed: $1 \cdot 2 \cdot 3 \cdots n > 1 \cdot 2 \cdot 2 \cdots 2$ by inspection.

(m)   Verify by substitution for $n = 1, 2, 3$. Use induction for $n \ge 3$.

1.9 Renumber the given fractions a_i/b_i in ascending order so that a_1/b_1 is the smallest and a_n/b_n is the largest. Write $x = a_1/b_1$ and $y = a_n/b_n$. Then $x \le a_i/b_i$ for $i = 2, \ldots, n$ so that $xb_i \le a_i$ for $i = 2, \ldots, n$. Add these $n-1$ inequalities to the equality $xb_1 = a_1$ to obtain

$$x \sum_{i=1}^{n} b_i \le \sum_{i=1}^{n} a_i .$$

This yields the left-hand version of the desired inequality. To get the right-hand version, add the equality $yb_n = a_n$ to the $n-1$ inequalities $yb_i \ge a_i$ for $i = 1, \ldots, n-1$.

1.10 Consider the algebraic formula

$$x^n - y^n = (x - y)(x^{n-1} + x^{n-2}y + \cdots + xy^{n-2} + y^{n-1}) . \tag{*}$$

(a)   The parenthetical sum on the right,

$$\sum_{k=1}^{n} x^{n-k}y^{k-1} ,$$

can be bounded above and below. Assume $x > y \ge 0$. Then $x^{n-k} > y^{n-k}$, and we can multiply through by y^{k-1} to get

$$x^{n-k}y^{k-1} > y^{n-1} .$$

Also, $x > y$ implies $x^{k-1} > y^{k-1}$ and we can multiply through by x^{n-k} to get

$$x^{n-1} > x^{n-k}y^{k-1} .$$

Therefore

$$y^{n-1} < x^{n-k}y^{k-1} < x^{n-1}$$

and summation over k gives

$$ny^{n-1} < \sum_{k=1}^{n} x^{n-k}y^{k-1} < nx^{n-1} .$$

Combine this with (*) to get the desired bound.

(b)   This bound follows from (*) because each of the n terms in the parenthetical sum on the right is less than or equal to $(\max\{|x|, |y|\})^{n-1}$ in absolute value.

1.11

(a)   The inequality holds trivially for $n = 1$. For the induction step use

$$\prod_{i=1}^{n+1}(1 + a_i) = (1 + a_{n+1})\prod_{i=1}^{n}(1 + a_i) = \prod_{i=1}^{n}(1 + a_i) + a_{n+1}\prod_{i=1}^{n}(1 + a_i)$$

$$\geq 1 + \sum_{i=1}^{n} a_i + a_{n+1}\prod_{i=1}^{n}(1 + a_i) \geq 1 + \sum_{i=1}^{n} a_i + a_{n+1} \ .$$

(b)   First check $n = 2$. Then

$$\prod_{i=1}^{n+1}(1 - a_i) = (1 - a_{n+1})\prod_{i=1}^{n}(1 - a_i) > (1 - a_{n+1})\left(1 - \sum_{i=1}^{n} a_i\right)$$

$$= 1 - a_{n+1} - \sum_{i=1}^{n} a_i + a_{n+1}\sum_{i=1}^{n} a_i > 1 - \sum_{i=1}^{n+1} a_i \ .$$

1.12 Let $\mathcal{P}(n)$ be $f_n < 2^n$. The cases $\mathcal{P}(1)$ and $\mathcal{P}(2)$ hold by inspection. The inequality

$$f_n = f_{n-1} + f_{n-2} < 2^{n-1} + 2^{n-2} = 3(2^{n-2}) < 4(2^{n-2}) = 2^n$$

shows that the truth of $\mathcal{P}(k)$ for $1 \leq k < n$ implies $\mathcal{P}(n)$. By the "strong" variation of the principle of induction, this is enough.

1.13 Use (1.19). The result is item 4.3.84 in Abramowitz and Stegun [1].

1.14 To get the result for A, sum the inequalities

$$\min a_n \leq a_n \leq \max a_n$$

for $1 \leq n \leq m$ and divide through by m.

1.16

(b)   For part (ii) we can begin by writing down the fact that

$$\sup_{x \in S} f(x) \geq f(x) \geq g(x) \quad \text{for all } x \in S.$$

This shows that $\sup_{x \in S} f(x)$ is an upper bound for $g(x)$ on S. By definition of $\sup_{x \in S} g(x)$ as the least upper bound of $g(x)$, we obtain the desired result. The proof for part (i) is similar. Note that a strict inequality $f(x) > g(x)$ would be blunted here as well.

(c)   For all $x \in S$ we have

$$f(x) \leq \sup_{x \in S} f(x) \quad \text{and} \quad g(x) \leq \sup_{x \in S} g(x) \ .$$

Adding these inequalities, we obtain

$$f(x) + g(x) = (f + g)(x) \leq \sup_{x \in S} f(x) + \sup_{x \in S} g(x) \quad \text{for all } x \in S.$$

So the number $\sup_{x \in S} f(x) + \sup_{x \in S} g(x)$ is an upper bound for $f(x) + g(x)$, and the desired inequality follows by definition of sup as the least upper bound.

1.17

(a)   By homogeneity we can set $b = 1$ and prove

$$\frac{1}{a} + 1 \leq a + \frac{1}{a^2} \qquad (a > 0)$$

instead. But this is easily put into the form $(a - 1)^2(a + 1) \geq 0$.

(b)   Set $b = 1$ again. The reduced inequality is equivalent to $(1 + \sqrt{a})(1 - \sqrt{a})^2 / \sqrt{a} \geq 0$.

(c)   Put $b = 1$ to get $a^5 - a^3 - a^2 + 1 = (a + 1)(a - 1)^2(a^2 + a + 1) \geq 0$.

1.18 [57] Without loss of generality, let $|a| \leq |b|$. Then $|a + b| \leq 2|b|$ and we have

$$|a + b|^p \leq 2^p |b|^p \leq 2^p (|a|^p + |b|^p) \,.$$

1.19 We look for a disk such that outside the disk the leading term of the polynomial dominates; that is, we seek ξ such that

$$|z| > \xi \quad \Longrightarrow \quad |a_0 z^n| > \left| \sum_{i=0}^{n-1} a_{n-i} z^i \right| \,.$$

First note that if $|z| \geq 1$ we have

$$\sum_{i=0}^{n-1} |z|^i \leq n|z|^{n-1}$$

and so

$$\left| \sum_{i=0}^{n-1} a_{n-i} z^i \right| \leq \sum_{i=0}^{n-1} |a_{n-i}| \, |z|^i \leq M \sum_{i=0}^{n-1} |z|^i \leq nM|z|^{n-1} \,,$$

where $M = \max |a_k|$. We now choose ξ as in (1.42) so that $|z| > \xi$ implies both $|z| > 1$ and $|z| > nM/|a_0|$. Then for $|z| > \xi$,

$$|a_0||z|^n - nM|z|^{n-1} > 0$$

and

$$|f(z)| = \left| a_0 z^n + \sum_{i=0}^{n-1} a_{n-i} z^i \right| \geq |a_0 z^n| - \left| \sum_{i=0}^{n-1} a_{n-i} z^i \right| \geq |a_0 z^n| - nM|z|^{n-1} > 0$$

by (1.19). Thus all zeros of $f(z)$ are in the disk $|z| \leq \xi$. (This argument is used in complex variable theory to prove the fundamental theorem of algebra via Rouché's theorem.) $\square$

2.1

(a)   Start with the obvious inequality $n! < (n + 1)^n$ and use the monotonicity of $\ln x$ to write $\ln(n!) < \ln(n + 1)^n$.

(b)   Write $\phi(\phi(x)) \leq \phi(f(x)) \leq f(f(x)) \leq f(\psi(x)) \leq \psi(\psi(x))$.

(c)   If f is increasing on $[a, b]$, then for any $x \in (a, b]$ we have

$$f(x) \geq \max_{u \in [a,x]} f(u) = \max_{u \in [a,x]} f(u) \cdot \frac{1}{x - a} \int_a^x du \geq \frac{1}{x - a} \int_a^x f(u)\, du \,.$$

This implies

$$\frac{1}{(x - a)^2} \left[(x - a)f(x) - \int_a^x f(u)\, du \right] \geq 0 \quad \text{or} \quad \frac{d}{dx} \left[\frac{1}{x - a} \int_a^x f(u)\, du \right] \geq 0$$

so that

$$\frac{1}{x - a} \int_a^x f(u)\, du \text{ is increasing on } (a, b] \,.$$

In particular,

$$\frac{1}{x-a}\int_a^x f(u)\,du \le \frac{1}{b-a}\int_a^b f(u)\,du \qquad (x \in (a,b)). \qquad (*)$$

Also, for any $x \in [a,b)$ we have

$$f(x) \le \min_{u\in[x,b]} f(u) = \min_{u\in[x,b]} f(u) \cdot \frac{1}{b-x}\int_x^b du \le \frac{1}{b-x}\int_x^b f(u)\,du.$$

This implies

$$\frac{1}{(b-x)^2}\left[-(b-x)f(x) + \int_x^b f(u)\,du\right] \ge 0 \quad \text{or} \quad \frac{d}{dx}\left[\frac{1}{b-x}\int_x^b f(u)\,du\right] \ge 0$$

so that

$$\frac{1}{b-x}\int_x^b f(u)\,du \text{ is increasing on } [a,b).$$

In particular,

$$\frac{1}{b-a}\int_a^b f(u)\,du \le \frac{1}{b-x}\int_x^b f(u)\,du \qquad (x \in [a,b)). \qquad (**)$$

Combining (*) and (**), we get (2.24).

2.2

(a) Fix ε and δ and a partition $a = x_0 < x_1 < \cdots < x_n = b$ with $x_i - x_{i-1} < \delta$ for all i. If $f(x)$ is not bounded on $[a,b]$, then it is not bounded on some subinterval $[x_{k-1}, x_k]$. Choose ξ_j for all $j \ne k$. There exists some ξ_k with $|f(\xi_k)|$ sufficiently large to contradict the assertion that

$$\left|\sum_{i=1}^n f(\xi_i)(x_i - x_{i-1}) - I\right| < \delta.$$

(b) Since $f(x)$ is integrable it is bounded, i.e., there exists M with $|f(x)| \le M$ on $[a,b]$. Let $\varepsilon > 0$ be given, and suppose $x_0 \in (a,b)$. Then

$$|F(x) - F(x_0)| = \left|\int_a^x f(t)\,dt - \int_a^{x_0} f(t)\,dt\right| = \left|\int_{x_0}^x f(t)\,dt\right| \le \left|\int_{x_0}^x |f(t)|\,dt\right|.$$

Hence

$$|F(x) - F(x_0)| \le M|x - x_0|$$

and we may choose $\delta = \varepsilon/M$.

(c) No. $f(x)$ is not bounded on $[0,1]$. However the integral exists as the improper integral

$$\lim_{\varepsilon \to 0^+}\int_\varepsilon^1 x^{-1/2}\,dx.$$

2.3

(a) $2^{-\alpha} \le I(\alpha,\beta) \le 1$.

(b) Use Jordan's inequality.

(c) For $s > b$,

$$\left|\int_0^\infty f(t)e^{-st}\,dt\right| \le \int_0^\infty |f(t)||e^{-st}|\,dt \le \int_0^\infty Ce^{bt}e^{-st}\,dt = \frac{C}{s-b}.$$

(d)   Use table lookup integrals to obtain

$$1 \le \frac{(2n-1)!!(2n+1)!!}{[(2n)!!]^2} \frac{\pi}{2} \le \frac{2n+1}{2n} \ .$$

As $n \to \infty$, the middle quantity is squeezed to 1.

(e)   Consider the alternating series

$$a_i = \int_{i\pi}^{(i+1)\pi} \frac{\sin x}{x}\, dx \ .$$

Since the alternating series has $|a_{i+1}| < |a_i|$ and $|a_i| \to 0$ as $i \to \infty$, the series converges. Write the series of positive terms

$$\int_0^\infty \frac{\sin x}{x}\, dx = (a_0 + a_1) + (a_2 + a_3) + \cdots > a_0 + a_1 \ .$$

Using Jordan's inequality on $[0, \pi/2]$ we have

$$\frac{\sin x}{x} \ge \frac{2}{\pi}, \quad \text{so} \quad \int_0^{\pi/2} \frac{\sin x}{x}\, dx \ge 1 \ .$$

Using symmetry and Fig. 2.3, on $[\pi/2, \pi]$ we have

$$\frac{\sin x}{x} \ge \frac{2}{x} - \frac{2}{\pi} \ge \frac{2}{\pi/2} - \frac{2}{\pi} \ ,$$

hence

$$\int_{\pi/2}^\pi \frac{\sin x}{x}\, dx \ge 1 \ .$$

Also

$$\int_\pi^{2\pi} \frac{\sin x}{x}\, dx \ge -2/3 \ .$$

(Sketch a box from π to 2π touching the curve $(\sin x)/x$ at $3\pi/2$.) Hence

$$\int_0^\infty \frac{\sin x}{x}\, dx > 4/3 \ .$$

Now regroup and get a_0 plus a series of negative terms,

$$\int_0^\infty \frac{\sin x}{x}\, dx = a_0 + (a_1 + a_2) + (a_3 + a_4) + \cdots < a_0 \ .$$

Form a left Riemann sum

$$\sum_{i=0}^2 f(x_i)\varDelta x > a_0 \ ,$$

where $\varDelta x = \pi/3$, $x_i = i\varDelta x$.

The integral from 0 to ∞ of the Fourier kernel $(\sin x)/x$ is computed using complex variables, with a crucial step invoking Jordan's lemma, which in turn uses Jordan's inequality. See [71]. The answer is $\pi/2$.

(f)   Assume the contrary and develop a contradiction based on Lemma 2.4.

(g)   Choose $\alpha < 1$ with $p + \alpha > 1$ (e.g., if $p \ge 1$ let $\alpha = 1/2$, if $0 < p < 1$ choose $1 - p < \alpha < 1$). For $x > 1$,

$$\ln x = \int_1^x (1/t)\, dt \le \int_1^x (1/t^\alpha)\, dt = (x^{1-\alpha} - 1)/(1 - \alpha)$$

hence $0 < \ln x / x^p \le (x^{1-\alpha-p} - x^{-p})/(1 - \alpha) \to 0$ as $x \to \infty$.

2.4 Integration of the inequality $0 \le g(t) \le 1$ over $[a, b]$ gives

$$0 \le \lambda \le b - a . \qquad (*)$$

To prove the right-hand part of Steffensen's inequality we note that $a \le a + \lambda \le b$. Then

$$\int_a^{a+\lambda} f(t)\, dt - \int_a^b f(t)g(t)\, dt = \int_a^{a+\lambda} f(t)\, dt - \int_a^{a+\lambda} f(t)g(t)\, dt - \int_{a+\lambda}^b f(t)g(t)\, dt$$

$$= \int_a^{a+\lambda} [1 - g(t)]f(t)\, dt - \int_{a+\lambda}^b f(t)g(t)\, dt$$

$$\ge f(a + \lambda) \int_a^{a+\lambda} [1 - g(t)]\, dt - \int_{a+\lambda}^b f(t)g(t)\, dt$$

since $t \le a + \lambda$ implies $f(t) \ge f(a + \lambda)$. Therefore

$$\int_a^{a+\lambda} f(t)\, dt - \int_a^b f(t)g(t)\, dt \ge f(a + \lambda) \left[\lambda - \int_a^{a+\lambda} g(t)\, dt \right] - \int_{a+\lambda}^b f(t)g(t)\, dt$$

$$= f(a + \lambda) \left[\int_a^b g(t)\, dt - \int_a^{a+\lambda} g(t)\, dt \right] - \int_{a+\lambda}^b f(t)g(t)\, dt$$

$$= f(a + \lambda) \int_{a+\lambda}^b g(t)\, dt - \int_{a+\lambda}^b f(t)g(t)\, dt$$

$$= \int_{a+\lambda}^b g(t)[f(a + \lambda) - f(t)]\, dt \ge 0$$

because $t \ge a + \lambda$ implies $f(t) \le f(a + \lambda)$. To prove the left-hand part, we write $(*)$ as $a \le b - \lambda \le b$ and obtain

$$\int_{b-\lambda}^b f(t)\, dt - \int_a^b f(t)g(t)\, dt = \int_{b-\lambda}^b f(t)\, dt - \int_{b-\lambda}^b f(t)g(t)\, dt - \int_a^{b-\lambda} f(t)g(t)\, dt$$

$$= \int_{b-\lambda}^b [1 - g(t)]f(t)\, dt - \int_a^{b-\lambda} f(t)g(t)\, dt$$

$$\le f(b - \lambda) \int_{b-\lambda}^b [1 - g(t)]\, dt - \int_a^{b-\lambda} f(t)g(t)\, dt$$

since $t \ge b - \lambda$ implies $f(t) \le f(b - \lambda)$. Therefore

$$\int_{b-\lambda}^b f(t)\, dt - \int_a^b f(t)g(t)\, dt \le f(b - \lambda) \left[\lambda - \int_{b-\lambda}^b g(t)\, dt \right] - \int_a^{b-\lambda} f(t)g(t)\, dt$$

$$= f(b - \lambda) \left[\int_a^b g(t)\, dt - \int_{b-\lambda}^b g(t)\, dt \right] - \int_a^{b-\lambda} f(t)g(t)\, dt$$

$$= f(b - \lambda) \int_a^{b-\lambda} g(t)\, dt - \int_a^{b-\lambda} f(t)g(t)\, dt$$

$$= \int_a^{b-\lambda} g(t)[f(b - \lambda) - f(t)]\, dt \le 0$$

since $t \le b - \lambda$ implies $f(t) \ge f(b - \lambda)$.

2.5

(a)   [78] First assume $f(b) = 0$. Since g is integrable on $[a, b]$ it is bounded, so we may choose a constant $c > 0$ with $g(x) + c \geq 0$ on $[a, b]$. Define the (continuous) function

$$G(\xi) = \int_a^\xi g(x)\, dx\,.$$

Let M denote the maximum and m the minimum of G on $[a, b]$. Let $\Delta x = (b - a)/n$ and $x_i = a + i\Delta x$ for $i = 0, \ldots, n$. Then

$$\int_a^b (g(x) + c)f(x)\, dx = \sum_{k=1}^n \int_{x_{k-1}}^{x_k} (g(x) + c)f(x)\, dx \leq \sum_{k=1}^n f(x_{k-1}) \int_{x_{k-1}}^{x_k} (g(x) + c)\, dx$$

$$= \sum_{k=1}^n f(x_{k-1}) \int_{x_{k-1}}^{x_k} g(x)\, dx + c \sum_{k=1}^n f(x_{k-1})(x_k - x_{k-1})\,.$$

In the last expression the first term can be rewritten

$$\sum_{k=1}^n f(x_{k-1})(G(x_k) - G(x_{k-1})) = \sum_{k=1}^n G(x_k)(f(x_{k-1}) - f(x_k))$$

$$\leq M \sum_{k=1}^n (f(x_{k-1}) - f(x_k)) = Mf(a)\,.$$

As $n \to \infty$ the second term approaches $c \int_a^b f(x)\, dx$. Taking limits, by Lemma 2.1,

$$\int_a^b (g(x) + c)f(x)\, dx \leq Mf(a) + c \int_a^b f(x)\, dx\,.$$

Similarly,

$$\int_a^b (g(x) + c)f(x)\, dx \geq mf(a) + c \int_a^b f(x)\, dx\,.$$

Now apply the intermediate value theorem to G. Finally, note that if $f(b) \neq 0$ we may redefine $f(x)$ to be 0 at $x = b$ without changing the integral $\int_a^b f(x)g(x)\, dx$.

(c)   If $f(x)$ is monotonic decreasing, replace $f(x)$ by $f(x) - f(b)$ in part (a). If $f(x)$ is monotonic increasing, replace $f(x)$ by $f(x) - f(a)$ in part (b).

(d)   Immediate from part (a).

2.6

(a)

$$\frac{n^3}{3} \leq \sum_{k=1}^n k^2 \leq \frac{(n + 1)^3 - 1}{3}\,.$$

(b)   Note that

$$\ln(n!) = \sum_{k=1}^n \ln k$$

and interpret this as a sum of rectangular areas (each of unit width).

(c)   Both trapezoids are bounded by the x-axis, and the lines $x = a, b$. The fourth side of the smaller trapezoid is formed by the line tangent to $y = 1/x$ at the midpoint $x = (a + b)/2$ of interval (a, b). The fourth side of the larger trapezoid is formed by the secant line connecting the points $x = a$, $y = 1/a$, and $x = b$, $y = 1/b$.

(d)   Use rectangles.

(e)   Use rectangles.

(f)   Draw a picture to see that

$$1 + \frac{1}{2^3} + \frac{1}{3^3} + \cdots + \frac{1}{n^3} < 1 + \frac{1}{2^3} + \sum_{k=3}^{\infty} \frac{1}{k^3} < 1 + \frac{1}{2^3} + \int_{2}^{\infty} \frac{dx}{x^3} = \frac{5}{4} \,.$$

(g)   We have

$$C_n - C_{n+1} = \ln(1 + 1/n) - 1/(n+1) > 0$$

by the logarithmic inequality, so C_n is decreasing. The lower bound of $1/2$ would require
that

$$\sum_{j=1}^{n} \frac{1}{j} - \frac{1}{2} > \ln n = \int_{1}^{n} \frac{dx}{x} \,.$$

To show that this is indeed the case, construct trapezoids to slightly overestimate the area
under $1/x$ as

$$A = \frac{1}{2} \sum_{j=1}^{n-1} \left(\frac{1}{j} + \frac{1}{j+1} \right) = \sum_{j=1}^{n} \frac{1}{j} - \frac{1}{2} - \frac{1}{2n} \,,$$

which makes it apparent that

$$\sum_{j=1}^{n} \frac{1}{j} - \frac{1}{2} > \ln n + \frac{1}{2n} > \ln n \,.$$

2.7

(a)   Define

$$f(x) = x - 1 - \ln x \,,$$

noting that $f(1) = 0$. We have $f'(x) = 1 - 1/x$, which is negative for $0 < x < 1$ and positive
for $x > 1$. The result, which is used in Chap. 3 to derive the weighted AM–GM inequality,
is item number 4.1.36 in Abramowitz and Stegun [1].

(b)   This is item 4.1.37 in Abramowitz and Stegun.

(c)   The function

$$f(x) = x^n - x + (n - 1)$$

has minimum value $f_{\min} = (n - 1)(1 - 1/n^{n/(n-1)})$ at $x = 1/n^{1/(n-1)}$.

(d)   Defining $f(x) = \sin x \tan x - 2 \ln(\sec x)$, we get $f'(x) = \sin x(\cos x - 1)^2 / \cos^2 x$.

(g)   This is item 4.2.31 in Abramowitz and Stegun [1]. After proving it, the reader might wish
to prove item 4.2.36, which is

$$e^x > (1 + x/y)^y > e^{xy/(x+y)} \qquad (x, y > 0) \,. \qquad\qquad (*)$$

The left-hand inequality in $(*)$, i.e., $e^{x/y} > 1 + x/y$, was stated in Example 2.7. To get the
right-hand inequality in $(*)$, we can start by using $e^{\xi} < (1 - \xi)^{-1}$ to write

$$e^{\xi/(\xi+\eta)} < (\xi + \eta)/\eta \quad \text{where} \quad \xi + \eta = 1 \,.$$

Now if $x + y = s$ where s may not be 1, then we still have $x/s + y/s = 1$ and can replace ξ
with x/s and η with y/s to get

$$e^{x/(x+y)} < (x + y)/y = 1 + x/y$$

where x and y are positive but otherwise unrestricted (the homogeneity idea). Finally, raise
both sides to the y power.

(h)   Differentiate the function $f(x) = \ln x/x$ to show that its maximum is attained at $x = e$, not
        at $x = \pi$, and hence that $f(e) > f(\pi)$.

(i)   Define $f(x) = x^a + (1 - x)^a$ for $0 \le x \le 1$, $0 < a < 1$. Check that $f(0) = f(1) = 1$, $f'(x) = 0$
        at $x = 1/2$ and $f(1/2) = 2^{1-a}$. Thus $1 \le x^a + (1 - x)^a \le 2^{1-a}$. Now substitute $x = s/(s + t)$.

(j)   Define $f(x) = x^b + (1 - x)^b$ for $0 \le x \le 1$, $b > 1$. Check that $f(0) = f(1) = 1$, $f'(x) = 0$ at
        $x = 1/2$ and $f(1/2) = 2^{1-b}$. Thus $2^{1-b} \le x^a + (1 - x)^a \le 1$. Now substitute $x = s/(s + t)$.

(k)   This can be obtained from the inequality $\ln x \le x - 1$. Putting $x \to x/a$ and rearranging
        gives $x \ge a[1 + \ln(x/a)] = a[\ln e + \ln(x/a)] = \ln[(ex/a)^a]$. Now raise e to the power of both
        sides. Equality holds iff $x = a$.

(l)   Write x^x as $e^{x \ln x}$. By monotonicity, the inequality $e^{x \ln x} \ge e^{x-1}$ is equivalent to $x \ln x \ge x - 1$.
        The latter inequality can be verified by defining $f(x) = x \ln x - (x - 1)$, which has $f(1) = 0$
        and $f'(x) = \ln x$. The result appears on p. 81 of Bullen [13].

(m)   Define $f(x) = \cos x - 1 + x^2/2$. Then $f(0) = 0$ and $f'(x) = x - \sin x \ge 0$ because $\sin x \le x$
        for $x \ge 0$.

(n)   Define $f(x) = \sin x - x + x^3/3!$. Then $f(0) = 0$ and $f'(x) = \cos x - 1 + x^2/2 \ge 0$ by the
        result of part (m).

2.8

(a)   Take $f(x) = \sin x$.

(b)   Take $f(x) = \tan^{-1} x$.

(c)   Take $f(x) = \sqrt{1 + x}$.

(d)   Take $f(x) = e^x$. More generally, we can take $f(x) = a^x$ where $a > 1$ and obtain

$$a^x(y - x) \ln a < a^y - a^x < a^y(y - x) \ln a \qquad (y > x).$$

(e)   Using $f(x) = (1 + x)^a$ in the mean value theorem, if $x > 0$ there exists $\xi \in (0, x)$ with

$$((1 + x)^a - 1)/x = a(1 + \xi)^{a-1} < a(1 + x)^{a-1},$$

and since $x > 0$ we have
$$(1 + x)^a - 1 < ax(1 + x)^{a-1}.$$

If $-1 < x < 0$ there exists ξ with $x < \xi < 0$ such that

$$((1 + x)^a - 1)/x = a(1 + \xi)^{a-1} > a(1 + x)^{a-1}$$

and since $x < 0$ we have
$$(1 + x)^a - 1 < ax(1 + x)^{a-1}.$$

(f)   By monotonicity we can take $\ln$ of both sides and see that this result is equivalent to the
        left-hand portion of the logarithmic inequality (2.11). It is item 4.2.34 in Abramowitz and
        Stegun [1].

(g)   Take $f(x) = \tan x$, $f'(x) = 1/\cos^2 x$.

(h)   Take $f(x) = x \ln x$.

2.9 Integrate both sides of the inequality $\sec^2 x \ge 1$, valid for $0 \le x < \pi/2$, over the interval $(0, x)$
to get $\tan x \ge x$ for $0 \le x < \pi/2$.

2.10

(a)   Write $h(x) = f(x)/g(x)$ with

$$f(x) = (1 + x)^a - 1 \quad \text{and} \quad g(x) = x.$$

Then $f'(x)/g'(x) = a(1 + x)^{a-1}$ is increasing (since its derivative is $a(a - 1)(1 + x)^{a-2} > 0$). Now use $f(0) = g(0) = 0$ and l'Hôpital's monotone rule (LMR) for the cases $x > 0$ giving $h(x) > h(0) = a$ and $x < 0$ giving $h(x) < h(0) = a$.

(b)

$$\frac{\ln \cosh x}{\ln((\sinh x)/x)}\Big|_{=0/0 \text{ at } x=0} \xrightarrow[\text{LMR}]{} \frac{x \tanh^2 x}{x - \tanh x}\Big|_{=0/0 \text{ at } x=0} \xrightarrow[\text{LMR}]{} 1 + 4x/(\sinh 2x)$$

which clearly decreases on $(0, \infty)$.

(c)   Check that $h(x) = \sin \pi x/(x(1 - x))$ is symmetric about $x = 1/2$, so we restrict to $(0, 1/2) = (a, b)$. Now $h(x) \to \pi$ as $x \to 0+$ by l'Hôpital's rule and $h(1/2) = 4$. To show h is monotonic,

$$\frac{\sin \pi x}{x(1 - x)}\Big|_{=0/0 \text{ at } x=0} \xrightarrow[\text{LMR}]{} \frac{\pi \cos \pi x}{1 - 2x}\Big|_{=0/0 \text{ at } x=1/2} \xrightarrow[\text{LMR}]{} (\pi^2 \sin \pi x)/2$$

is increasing on $(0, 1/2)$.

(d) $h(x) = \sin x/x$ on $(0, \pi/2]$ extends to $[0, \pi/2]$ with $h(0) = 1$ by l'Hôpital's rule, $h(\pi/2) = 2/\pi$. Now

$$\frac{\sin x}{x}\Big|_{=0/0 \text{ at } x=0} \xrightarrow[\text{LMR}]{} \frac{\cos x}{1}$$

is strictly decreasing hence $1 > \sin x/x > 2/\pi$ on $(0, \pi/2)$.

2.11

(a)   This is item 4.3.86 in Abramowitz and Stegun [1].

(b)   This is item 4.1.38 in Abramowitz and Stegun [1].

(c)   Apply $e^x \geq 1 + x$ (the weak version of a result stated in Example 2.7 and cited as item 4.2.30 in Abramowitz and Stegun [1]). Thus

$$1 + a_n \leq \exp(a_n) \qquad (n = 1, \ldots, N)$$

and

$$\prod_{n=1}^{N}(1 + a_n) \leq \prod_{n=1}^{N} \exp(a_n) = \exp\left(\sum_{n=1}^{N} a_n\right).$$

(d)   This is item 4.2.35 in Abramowitz and Stegun [1].

(e)   The left-hand inequality is $e^x > 1 + x$, stated in Example 2.7 and cited as item 4.2.30 in Abramowitz and Stegun [1]. The right-hand inequality is equivalent to $e^x < (1 - x)^{-1}$, obtained in Problem 2.7.

2.12 Substitute $x = a/b > 1$ and use either differentiation or Corollary 2.4 to establish that

$$n(x - 1) < x^n - 1 < nx^{n-1}(x - 1).$$

Next, suppose that $a^n = b^n$ with $a \neq b$; then $b^{n-1} < 0 < a^{n-1}$ (a contradiction, because b assumed positive).

2.13 Between each x_i and x_{i+1} there is a point where $g'(x)$ vanishes. Between every two adjacent such points there is a point where $g''(x)$ vanishes. Continue the pattern until reaching a point ξ where $g^{(n)}(x)$ vanishes.

2.14 Let A be dense in B, and suppose $x_n \to x$ where $x \in B$ and x_n is a sequence of points in A. By hypothesis $f(x_n) \leq g(x_n)$ for all n; hence, by Lemma 2.1 we know that $\lim f(x_n) \leq \lim g(x_n)$ and by continuity (Theorem 2.1) the result is proved. Also note that the rationals are dense in the reals.

2.15 No; think of two functions $f(x)$ and $g(x)$ (e.g., two straight lines defined on some interval of the x-axis) such that $f(x)$ is greater than $g(x)$ but the slope of $f(x)$ is less than that of $g(x)$.

2.16 Find the minimum of the function

$$z = \tfrac{1}{2}(x^n + y^n) \text{ subject to the condition } x + y = s .$$

The solution of the Lagrange multiplier system

$$\tfrac{1}{2}nx^{n-1} - \lambda = 0 , \quad \tfrac{1}{2}ny^{n-1} - \lambda = 0, \quad x + y = s ,$$

is $x = y = s/2$, so the minimum value of z is $(s/2)^n$, which is the right member of the desired inequality.

3.1

(a) $g(f(x)) = (x^{p-1})^{q-1} = x^{(p-1)(q-1)} = x^1 = x.$

(b)   In (3.13), set $w = a\varepsilon^{1/p}$ and $h = b\varepsilon^{(1-q)/q}$.

(c)   Put $y = f(x) = \ln(x+1)$, noting that its inverse function is $x = g(y) = e^y - 1$. Use the integral

$$\int \ln(x+1)\, dx = (x+1)\ln(x+1) - x + C .$$

(d)   For any $\lambda, \mu \geq 0$ such that $\lambda + \mu = 1$ and $x, y > 0$, we have

$$\ln(\lambda x + \mu y) \geq \lambda \ln x + \mu \ln y .$$

Therefore

$$\ln\left(\frac{1}{p}a^p + \frac{1}{q}b^q\right) \geq \frac{1}{p}\ln(a^p) + \frac{1}{q}\ln(b^q) = \ln a + \ln b = \ln(ab) .$$

Finally, apply the monotonicity of the log function.

3.2

(a)   This is AM–GM:

$$\frac{a^4 + b^3 + c^4 + d^4}{4} \geq \sqrt[4]{a^4 b^4 c^4 d^4} .$$

(b)   Multiply the three inequalities

$$\frac{a+b}{2} \geq \sqrt{ab} , \quad \frac{b+c}{2} \geq \sqrt{bc} , \quad \frac{c+a}{2} \geq \sqrt{ca} .$$

(c)   Apply AM–GM to each of the two factors on the left:

$$\frac{ab + bc + ca}{3} \cdot \frac{a^4 + b^4 + c^4}{3} \geq \sqrt[3]{ab \cdot bc \cdot ca} \cdot \sqrt[3]{a^4 \cdot b^4 \cdot c^4} = a^2 b^2 c^2 .$$

(d)   Write

$$\frac{a+b}{2} \geq \sqrt{ab} , \qquad \frac{\frac{1}{a} + \frac{1}{b}}{2} \geq \sqrt{\frac{1}{a} \cdot \frac{1}{b}} ,$$

and add.

(e)   If $a = b$ then equality holds as $0 = 0$. If $a \neq b$, then we can divide through by $(a - b)^2$ and get the equivalent inequality

$$(a + b)^2 \geq 4ab$$

which is equivalent to AM–GM.

(f)   Apply AM–GM to the numbers ad and bc to get

$$ad + bc \leq 2\sqrt{abcd} .$$

Now add $ab + cd$ to both sides, factor as

$$(a + c)(b + d) \geq (\sqrt{ab} + \sqrt{cd})^2 ,$$

and take square roots.

(g)   Apply AM–GM to the numbers $a/\sqrt{b}$ and $b/\sqrt{a}$.

(h)   Multiply the results

$$\frac{ab + cd}{2} \geq \sqrt{abcd} , \qquad \frac{ac + bd}{2} \geq \sqrt{acbd} .$$

(i)   Multiply the results

$$\frac{a^2b + b^2c + c^2a}{3} \geq (a^2b \cdot b^2c \cdot c^2a)^{1/3} , \qquad \frac{ab^2 + bc^2 + ca^2}{3} \geq (ab^2 \cdot bc^2 \cdot ca^2)^{1/3} .$$

(j)   Apply AM–GM as

$$\frac{1}{2}\left(\frac{bc}{a} + \frac{ac}{b}\right) + \frac{1}{2}\left(\frac{ac}{b} + \frac{ab}{c}\right) + \frac{1}{2}\left(\frac{bc}{a} + \frac{ab}{c}\right) \geq \sqrt{\frac{bc}{a} \cdot \frac{ac}{b}} + \sqrt{\frac{ac}{b} \cdot \frac{ab}{c}} + \sqrt{\frac{bc}{a} \cdot \frac{ab}{c}} .$$

3.3

(a)   The summation formula

$$\sum_{k=1}^{n} k = \frac{n(n+1)}{2}$$

is useful in this problem. Putting $a_k = k$ for $k = 1, \ldots, n$, we apply AM–GM as

$$\sqrt[n]{n!} < \frac{1}{n}\sum_{k=1}^{n} k = \frac{1}{n} \cdot \frac{n(n+1)}{2} = \frac{n+1}{2} .$$

Next,

$$[(2n)!!]^{1/n} < \frac{(2n) + (2n-2) + \cdots + 4 + 2}{n} = \frac{2\sum_{i=1}^{n} i}{n} = n + 1 .$$

(b)   Apply AM–GM with $a_i = b$ for $i = 1, \ldots, n-1$, and $a_n = a$.

(c)   [29] Put $a_i = i^3$ for $i = 1, \ldots, n$.

(d)   [29] Put $(a_i) = \dfrac{1}{1 \cdot 2}, \dfrac{1}{2 \cdot 3}, \ldots, \dfrac{1}{n(n+1)}$ to get

$$\frac{1}{n}\sum_{k=1}^{n} \frac{1}{k(k+1)} > \left[\frac{1}{(n!)^2(n+1)}\right]^{1/n} . \qquad (*)$$

The sum on the left telescopes,

$$\sum_{k=1}^{n} \frac{1}{k(k+1)} = \sum_{k=1}^{n}\left(\frac{1}{k} - \frac{1}{k+1}\right) = 1 - \frac{1}{n+1} = \frac{n}{n+1} ,$$

so the left-hand side of $(*)$ reduces to $1/(n+1)$. Now solve for n!

(e)   AM–GM looks like

$$\frac{1}{n}\left(\frac{x_1}{x_2} + \frac{x_2}{x_3} + \cdots + \frac{x_{n-1}}{x_n} + \frac{x_n}{x_1}\right) \geq \left(\frac{x_1}{x_2} \cdot \frac{x_2}{x_3} \cdots \frac{x_{n-1}}{x_n} \cdot \frac{x_n}{x_1}\right)^{1/n} = 1 .$$

(f)   AM–GM looks like

$$\frac{1}{n}(x_1^n + x_2^n + \cdots + x_n^n) \geq (x_1^n x_2^n \cdots x_n^n)^{1/n} = x_1 \cdots x_n \,.$$

(g)   Apply AM–GM to the $n + 1$ numbers

$$1 \quad \text{and} \quad \underbrace{a^{n+1}, \ldots, a^{n+1}}_{n \text{ times}} \,.$$

This result is a special case of

$$x^{n+1} + n y^{n+1} \geq (n + 1) x y^n \qquad (x, y \geq 0)$$

which Pachpatte [70] uses to derive Hardy's inequality. (Apply AM–GM to x^{n+1} along with the n numbers y^{n+1}.)

(h)   Apply AM–GM to the n numbers

$$a_n \quad \text{and} \quad \underbrace{\sqrt[n-1]{a_1 \cdots a_{n-1}}, \ \ldots, \ \sqrt[n-1]{a_1 \cdots a_{n-1}}}_{n-1 \text{ times}} \,.$$

(i)   Let $a_1 a_2 \cdots a_N = 1$ where each $a_i > 0$. By AM–GM,

$$\frac{1}{N}(a_1 + a_2 + \cdots + a_N) \geq (a_1 a_2 \cdots a_N)^{1/N} = 1$$

so $a_1 + a_2 + \cdots + a_N \geq N$.

(j)   Multiply by 1 and then use AM–GM. Write

$$[(1 - 1/n)^n \cdot 1]^{1/(n+1)} < \frac{1}{n+1}[1 + n(1 - 1/n)] = n/(n+1) = 1 - 1/(n+1)$$

and raise both sides to the $n + 1$ power. The same trick can be used with weighted AM–GM to show that if $\xi > 0$ and $0 < m < n$, then

$$(1 + \xi/m)^m < (1 + \xi/n)^n \,.$$

Apply the inequality to the numbers $1 + \xi/m$ and 1 with the respective weights m/n and $(n - m)/n$:

$$(1 + \xi/m)^{m/n} \cdot 1^{(n-m)/n} < (m/n)(1 + \xi/m) + [(n - m)/n] \cdot 1 = 1 + \xi/n$$

and then raise both sides to the nth power [34].

(k)   Use AM–GM twice:

$$\sum_{n=1}^{N} a_n^{-m} \geq N \left(\prod_{n=1}^{N} a_n^{-m} \right)^{1/N} = N \left[\frac{1}{(\prod_{n=1}^{N} a_n)^{1/N}} \right]^m \geq N \left[\frac{1}{(1/N) \sum_{n=1}^{N} a_n} \right]^m \,.$$

(l)   [54] By AM–GM we have

$$r(n - r) \leq (n/2)^2 \qquad (1 \leq r \leq n - 1) \,.$$

Multiplying, we get

$$[(n - 1)!]^2 \leq [(n/2)^2]^{n-1} \,.$$

Taking the square root of both sides and multiplying through by n, we get the result.

3.4

(a) Let the rectangle have length L, width W, perimeter P. Then

$$P/4 = (L + W)/2 \geq (LW)^{1/2} .$$

Equality occurs iff $L = W$.

(b) The force of repulsion is given by Coulomb's law

$$F = k \frac{q(Q - q)}{R^2}$$

where k is a constant and R is the separation distance (held constant). So we seek the maximum value of the product $q(Q - q)$ for a given Q. By AM–GM,

$$q(Q - q) \leq \left(\frac{q + (Q - q)}{2} \right)^2 = Q^2/4 .$$

Equality holds when $q = Q - q$, in other words when $q = Q/2$.

3.5

(a) The result (3.7) holds for $n = 2$ because $(\sqrt{a_1} - \sqrt{a_2})^2 \geq 0$. Equality holds iff $a_1 = a_2$. Supposing (3.7) holds for some $n \geq 2$, we show that it holds for $2n$. We first use the induction hypothesis, then the established case for $n = 2$:

$$\frac{a_1 + a_2 + \cdots + a_{2n}}{2n} = \frac{\dfrac{a_1 + a_2 + \cdots + a_n}{n} + \dfrac{a_{n+1} + a_{n+2} + \cdots + a_{2n}}{n}}{2}$$

$$\geq \frac{(a_1 a_2 \cdots a_n)^{1/n} + (a_{n+1} a_{n+2} \cdots a_{2n})^{1/n}}{2}$$

$$\geq [(a_1 a_2 \cdots a_n)^{1/n} (a_{n+1} a_{n+2} \cdots a_{2n})^{1/n}]^{1/2} = (a_1 a_2 \cdots a_{2n})^{1/2n} .$$

Equality holds iff $a_1 = a_2 = \cdots = a_n$ and $a_{n+1} = a_{n+2} = \cdots = a_{2n}$ and

$$(a_1 a_2 \cdots a_n)^{1/n} = (a_{n+1} a_{n+2} \cdots a_{2n})^{1/n} .$$

So equality holds iff $a_1 = a_2 = \cdots = a_{2n}$. Finally, we show that if (3.7) holds for some $n \geq 4$, then it holds for $n - 1$. Denote by α the arithmetic mean of $a_1, a_2, \ldots, a_{n-1}$. Since the mean of the set $\{a_1, a_2, \ldots, a_{n-1}, \alpha\}$ is also equal to α, we have

$$\alpha = \frac{a_1 + a_2 + \cdots + a_{n-1} + \alpha}{n}$$

hence by the induction hypothesis

$$\alpha \geq (a_1 a_2 \cdots a_{n-1} \cdot \alpha)^{1/n} .$$

We get $\alpha \geq (a_1 a_2 \cdots a_{n-1})^{1/(n-1)}$ with equality iff $a_1 = a_2 = \cdots = a_{n-1}$.

(b) We rephrase the problem as follows:

Maximize the function $f(x_1, x_2, \ldots, x_n) = (x_1 \cdot x_2 \cdots \cdot x_n)^{1/n}$ subject to the constraint that the sum of all the x_i is equal to some given number C.

This is a Lagrange multiplier problem with the single constraint

$$g(x_1, x_2, \ldots, x_n) = x_1 + x_2 + \cdots + x_n - C = 0 ,$$

hence the Lagrangian function is of the form

$$(x_1 \cdot x_2 \cdot \cdots \cdot x_n)^{1/n} + \lambda(x_1 + x_2 + \cdots + x_n - C) .$$

Differentiating it with respect to x_1 (treating λ as a constant) and setting the result to zero, we obtain

$$\frac{1}{n}(x_1 \cdot x_2 \cdot \cdots \cdot x_n)^{\frac{1}{n}-1}(x_2 \cdots x_n) + \lambda = \frac{1}{n}\frac{f(x_1, x_2, \ldots, x_n)}{x_1} + \lambda = 0 .$$

Similarly we obtain

$$\frac{1}{n}\frac{f(x_1, x_2, \ldots, x_n)}{x_i} + \lambda = 0 \qquad (i = 2, \ldots, n)$$

after differentiation with respect to the remaining x_i. These n equations yield

$$f(x_1, x_2, \ldots, x_n) = -n\lambda x_1 = -n\lambda x_2 = \cdots = -n\lambda x_n ,$$

hence $x_1 = x_2 = \cdots = x_n = C/n$ by the constraint equation. Evaluation of f at this point gives C/n, hence

$$(x_1 \cdot x_2 \cdot \cdots \cdot x_n)^{1/n} \le \frac{C}{n} = \frac{x_1 + x_2 + \cdots + x_n}{n} .$$

3.6 The case $n = 1$ is trivial. For $n \ge 2$ consider a_n as a variable by defining for $x > 0$,

$$f(x) = \frac{\delta_1 a_1 + \cdots + \delta_{n-1} a_{n-1} + \delta_n x}{a_1^{\delta_1} \cdots a_{n-1}^{\delta_{n-1}} x^{\delta_n}} = \frac{s_{n-1} + \delta_n x}{p_{n-1} x^{\delta_n}} , \text{ say.}$$

Show that $f'(x) = 0$ at $x_m = s_{n-1}/(1 - \delta_n)$ and

$$f''(x_m) = (\delta_n/p_{n-1})x_m^{-\delta_n-1}(1 - \delta_n) > 0 ;$$

hence $f(x)$ has its minimum at x_m where

$$f(x_m) = \frac{\{[\delta_1/(1 - \delta_n)]a_1 + \cdots + [\delta_{n-1}/(1 - \delta_n)]a_{n-1}\}^{1-\delta_n}}{a_1^{\delta_1} \cdots a_{n-1}^{\delta_{n-1}}} .$$

The weights $\delta_1/(1 - \delta_n), \ldots, \delta_{n-1}/(1 - \delta_n)$ add up to 1, so inductively if $a_1, \ldots, a_{n-1}$ are not all equal, then

$$[\delta_1/(1 - \delta_n)]a_1 + \cdots + [\delta_{n-1}/(1 - \delta_n)]a_{n-1} > a_1^{\delta_1/(1-\delta_n)} \cdots a_{n-1}^{\delta_{n-1}/(1-\delta_n)}$$

so

$$f(x) \ge f(x_m) > \frac{[a_1^{\delta_1/(1-\delta_n)} \cdots a_{n-1}^{\delta_{n-1}/(1-\delta_n)}]^{1-\delta_n}}{a_1^{\delta_1} \cdots a_{n-1}^{\delta_{n-1}}} = 1 .$$

If $a_1 = \cdots = a_{n-1}$, then

$$x_m = [\delta_1/(1 - \delta_n)]a_1 + \cdots + [\delta_{n-1}/(1 - \delta_n)]a_{n-1} = a_1$$

and $f(x_m) = 1$. For any other choice of x we have $f(x) > f(x_m) = 1$.

3.7

(a)   We have

$$\lim_{t \to 0} \ln g(t) = \lim_{t \to 0} \left[\frac{1}{t} \ln \sum_{i=1}^{n} \delta_i x_i^t\right] = \lim_{t \to 0} \frac{\sum_{i=1}^{n} \delta_i x_i^t \ln x_i}{\sum_{i=1}^{n} \delta_i x_i^t} = \sum_{i=1}^{n} \delta_i \ln x_i$$

by l'Hôpital's rule and the fact that $\sum_{i=1}^{n} \delta_i = 1$. Hence

$$g(t) \to \prod_{i=1}^{n} x_i^{\delta_i} \qquad (t \to 0) .$$

(b)

$$\ln g(t) = \left(\frac{1}{t} \ln \sum_{i=1}^{n} \delta_i x_i^t \right)\Big|_{=0/0 \text{ at } t=0} \xrightarrow[\text{LMR}]{} \left(\sum_{i=1}^{n} \delta_i x_i^t \ln x_i \right) \Big/ \left(\sum_{i=1}^{n} \delta_i x_i^t \right) .$$

The last expression is increasing since its derivative, using the quotient rule, has numerator

$$\left(\sum_{i=1}^{n} \delta_i x_i^t \right)\left(\sum_{i=1}^{n} \delta_i x_i^t \ln^2 x_i \right) - \left(\sum_{i=1}^{n} \delta_i x_i^t \ln x_i \right)^2$$

which is nonnegative because

$$\left(\sum_{i=1}^{n} \delta_i x_i^t \ln x_i \right)^2 = \left(\sum_{i=1}^{n} \delta_i^{1/2} x_i^{t/2} \delta_i^{1/2} x_i^{t/2} \ln x_i \right)^2 \le \left(\sum_{i=1}^{n} \delta_i x_i^t \right)\left(\sum_{i=1}^{n} \delta_i x_i^t \ln^2 x_i \right)$$

by the Cauchy–Schwarz inequality (3.25). This applies to g on $(0, \infty)$ or $(-\infty, 0)$.

3.8 Choose a partition $x_0 = a$, $x_1 = a + \Delta x, \ldots, x_n = b$ where $\Delta x = (b-a)/n$. Form Riemann sums approximating each term with $a_1 = f(x_1), \ldots, a_n = f(x_n)$ so that

$$\frac{a_1 + \cdots + a_n}{n} = \frac{[f(x_1) + \cdots + f(x_n)]\Delta x}{b-a} \to \frac{1}{b-a} \int_a^b f(x)\, dx ,$$

$$\left(\frac{a_1^{-1} + \cdots + a_n^{-1}}{n} \right)^{-1} \to \frac{b-a}{\int_a^b (1/f(x))\, dx} ,$$

and

$$(a_1 \cdots a_n)^{1/n} = \exp\left[\frac{\ln a_1 + \cdots + \ln a_n}{n} \right]$$

$$= \exp\left[\frac{\Delta x(\ln(f(x_1)) + \cdots + \ln(f(x_n)))}{b-a} \right]$$

$$\to \exp\left[\frac{1}{b-a} \int_a^b \ln f(x)\, dx \right] \quad \text{as } n \to \infty .$$

Use the previous problem (c) with each $\delta_i = 1/n$ and Lemma 2.1.

3.9 To simplify notation, denote $y_i = x_i^t$ and $s = \sum_{i=1}^{n} y_i$. We first note that

$$s^s = s^{\sum_{i=1}^{n} y_i} = \prod_{i=1}^{n} s^{y_i} \ge \prod_{i=1}^{n} y_i^{y_i} \quad \text{and hence} \quad \ln s \ge \frac{1}{s} \sum_{i=1}^{n} y_i \ln y_i .$$

But then

$$\frac{d}{dt} \ln h(t) = \frac{1}{t^2}\left[\frac{1}{s} \sum_{i=1}^{n} y_i \ln y_i - \ln s \right] \le 0 .$$

3.10

(a)   Rearrange the AM–HM inequality.

(b)   Use the AM–QM inequality in the form $a^2 + b^2 \geq \frac{1}{2}(a+b)^2$:

$$a^4 + b^4 \geq \frac{1}{2}\left[a^2+b^2\right]^2 \geq \frac{1}{2}\left[\frac{1}{2}(a+b)^2\right]^2 \geq \frac{1}{2}\left[\frac{1}{2}\right]^2 = \frac{1}{8}.$$

(c)   Apply the GM–HM inequality three times as

$$\sqrt{\frac{a}{b+c}}\cdot 1 \geq \frac{2}{1+\frac{b+c}{a}} = \frac{2a}{a+b+c},$$

$$\sqrt{\frac{b}{c+a}}\cdot 1 \geq \frac{2}{1+\frac{c+a}{b}} = \frac{2b}{b+c+a},$$

$$\sqrt{\frac{c}{a+b}}\cdot 1 \geq \frac{2}{1+\frac{a+b}{c}} = \frac{2c}{c+a+b},$$

and add the results. The condition for equality

$$1 = \frac{a}{b+c} = \frac{b}{c+a} = \frac{c}{a+b}$$

cannot hold, so strict inequality holds.

(d)   Apply the QM–AM inequality to write

$$\sqrt{\frac{\left(x_1+\frac{1}{x_1}\right)^2 + \cdots + \left(x_n+\frac{1}{x_n}\right)^2}{n}} \geq \frac{\left(x_1+\frac{1}{x_1}\right) + \cdots + \left(x_n+\frac{1}{x_n}\right)}{n} = \frac{1 + \left(\frac{1}{x_1}+\cdots+\frac{1}{x_n}\right)}{n}$$

However, by the AM–HM inequality we have

$$\frac{x_1 + \cdots + x_n}{n} \geq \frac{n}{\frac{1}{x_1}+\cdots+\frac{1}{x_n}}$$

which yields

$$\frac{1}{x_1} + \cdots + \frac{1}{x_n} \geq n^2 \quad \text{if } x_1 + \cdots + x_n = 1.$$

3.11 Ptak [76] gives a proof using AM–GM as follows. From the given a_i, form a new set of numbers

$$b_i = a_i/\sqrt{a_1 a_m} \qquad (i = 1, 2, \ldots, m).$$

Like the a_i, these are positive and satisfy $b_1 < b_2 < \ldots < b_m$; moreover

$$b_m = a_m/\sqrt{a_1 a_m} = \sqrt{a_1 a_m}/a_1 = 1/b_1$$

so that

$$b_i \leq 1/b_1 \qquad (i = 1, 2, \ldots, m).$$

Manipulate this to get

$$b_i - b_1 \leq \frac{b_i - b_1}{b_i b_1} = \frac{1}{b_1} - \frac{1}{b_i} \quad \text{or} \quad b_i + \frac{1}{b_i} \leq b_1 + \frac{1}{b_1} \qquad (i = 1, 2, \ldots, m).$$

Hence

$$\sum_{i=1}^{m} \lambda_i b_i + \sum_{i=1}^{m} \frac{\lambda_i}{b_i} \leq \left(b_1 + \frac{1}{b_1}\right) \sum_{i=1}^{m} \lambda_i$$

or

$$\frac{1}{2}\left[\sum_{i=1}^{m} \lambda_i b_i + \sum_{i=1}^{m} \frac{\lambda_i}{b_i}\right] \leq \frac{b_1 + b_m}{2} .$$

By AM–GM then,

$$\left(\sum_{i=1}^{m} \lambda_i b_i\right)\left(\sum_{i=1}^{m} \frac{\lambda_i}{b_i}\right) \leq \left(\frac{b_1 + b_m}{2}\right)^2 .$$

When rewritten in terms of the a_i, this is the desired result.

3.12 If (3.42) holds, then

$$c = \left(\frac{\sum_{i=1}^{m} |b_i|^q}{\sum_{i=1}^{m} |a_i|^p}\right)^{1/q} .$$

If (3.43) holds, then for each i we have

$$\frac{|b_i|^q}{\sum_{i=1}^{m} |b_i|^q} = \frac{(c|a_i|^{p-1})^q}{\sum_{i=1}^{m} (c|a_i|^{p-1})^q} = \frac{|a_i|^p}{\sum_{i=1}^{m} |a_i|^p} .$$

3.13 We prove Hölder's inequality

$$\sum_{i=1}^{n} a_i x_i \leq \left(\sum_{i=1}^{n} a_i^p\right)^{1/p}\left(\sum_{i=1}^{n} x_i^q\right)^{1/q} \qquad (a_i, x_i \geq 0)$$

by finding the minimum of the function

$$u = \left(\sum_{i=1}^{n} a_i^p\right)^{1/p}\left(\sum_{i=1}^{n} x_i^q\right)^{1/q} \text{ subject to the condition } \sum_{i=1}^{n} a_i x_i = A .$$

The Lagrange multiplier system consists of

$$\left(\sum_{i=1}^{n} a_i^p\right)^{\frac{1}{p}} \cdot \frac{1}{q} \cdot \left(\sum_{i=1}^{n} x_i^q\right)^{\frac{1}{q}-1} \cdot q x_i^{q-1} - \lambda a_i = 0$$

or

$$\left(\sum_{i=1}^{n} a_i^p\right)^{\frac{1}{p}}\left(\sum_{i=1}^{n} x_i^q\right)^{\frac{1}{q}-1} \frac{x_i^{q-1}}{a_i} = \lambda \qquad (i = 1, \ldots, n)$$

and the condition stated. Thus

$$\frac{x_1^{q-1}}{a_1} = \cdots = \frac{x_n^{q-1}}{a_n} \quad \text{so that} \quad \frac{x_1}{a_1^{1/(q-1)}} = \cdots = \frac{x_n}{a_n^{1/(q-1)}}$$

and hence

$$x_i = \frac{a_i^{1/(q-1)}}{a_1^{1/(q-1)}} x_1 \qquad (i = 1, \ldots, n) .$$

The condition gives

$$A = \frac{x_1}{a_1^{1/(q-1)}} \sum_{i=1}^{n} a_i a_i^{1/(q-1)} = \frac{x_1}{a_1^{1/(q-1)}} \sum_{i=1}^{n} a_i^p \quad \text{since} \quad 1 + \frac{1}{q-1} = \frac{q}{q-1} = p .$$

Therefore

$$x_1 = \frac{a_1^{1/(q-1)} A}{\sum_{i=1}^n a_i^p} \qquad \text{and} \qquad x_i = \frac{a_i^{1/(q-1)}}{\sum_{i=1}^n a_i^p} A \qquad (i = 1, \ldots, n).$$

Evaluation of u at these x values gives

$$u_{\min} = \left(\sum_{i=1}^n a_i^p\right)^{1/p} \left(\sum_{i=1}^n \left[\frac{a_i^{1/(q-1)}}{\sum_{i=1}^n a_i^p} A\right]^q\right)^{1/q} = \left(\sum_{i=1}^n a_i^p\right)^{1/p} \left(\sum_{i=1}^n a_i^{q/(q-1)}\right)^{1/q} \frac{A}{\sum_{i=1}^n a_i^p}$$

$$= \left(\sum_{i=1}^n a_i^p\right)^{1/p} \left(\sum_{i=1}^n a_i^p\right)^{1/q} \frac{A}{\sum_{i=1}^n a_i^p} = \left(\sum_{i=1}^n a_i^p\right)^{\frac{1}{p}+\frac{1}{q}-1} A = A.$$

3.14 Let $a = x_0$, $x_1 = x_0 + \Delta x$, ..., $x_m = b$ where $\Delta x = (b-a)/m$. Calling $a_i = f(x_i)$ and $b_i = g(x_i)$, we have

$$\sum_{i=1}^m |a_i b_i| \Delta x \to \int_a^b |f(x)g(x)| \, dx,$$

$$\left(\sum_{i=1}^m |a_i|^p \Delta x\right)^{1/p} \to \left(\int_a^b |f(x)| \, dx\right)^{1/p}, \qquad \left(\sum_{i=1}^m |b_i|^q \Delta x\right)^{1/q} \to \left(\int_a^b |g(x)| \, dx\right)^{1/q},$$

as $m \to \infty$. Now use (3.11) and Lemma 2.1.

3.15 Write

$$\sum_{k=1}^n |a_k| = \sum_{k=1}^n |a_k \cdot 1| \le \left(\sum_{k=1}^n |a_k|^p\right)^{1/p} \left(\sum_{k=1}^n 1^q\right)^{1/q} = n^{1/q} \left(\sum_{k=1}^n |a_k|^p\right)^{1/p},$$

then raise both sides to the p power and use the fact that $p/q = p - 1$.

3.16 The case $n = 2$ is Minkowski's inequality, so it holds. We now assume that (3.44) holds for $n = N$ and show that it also holds for $n = N + 1$. Writing

$$b_i = \sum_{k=1}^N a_i^{(k)}$$

and using Minkowski's inequality, we have

$$\left[\sum_{i=1}^m \left(\sum_{k=1}^{N+1} a_i^{(k)}\right)^p\right]^{1/p} = \left[\sum_{i=1}^m \left(b_i + a_i^{(N+1)}\right)^p\right]^{1/p} \le \left[\sum_{i=1}^m \left(\sum_{k=1}^N a_i^{(k)}\right)^p\right]^{1/p} + \left[\sum_{i=1}^m \left(a_i^{(N+1)}\right)^p\right]^{1/p}.$$

Hence by the induction hypothesis

$$\left[\sum_{i=1}^m \left(\sum_{k=1}^{N+1} a_i^{(k)}\right)^p\right]^{1/p} \le \sum_{k=1}^N \left(\sum_{i=1}^m (a_i^{(k)})^p\right)^{1/p} + \left(\sum_{i=1}^m (a_i^{(N+1)})^p\right)^{1/p} = \sum_{k=1}^{N+1} \left(\sum_{i=1}^m (a_i^{(k)})^p\right)^{1/p}.$$

3.18 Use

$$\int_a^b |f(x) + g(x)|^2 \, dx \le \int_a^b |f(x) + g(x)|^2 \, dx + \int_a^b |f(x) - g(x)|^2 \, dx$$

$$= \int_a^b [f(x) + g(x)]^2 \, dx + \int_a^b [f(x) - g(x)]^2 \, dx$$

$$= 2 \int_a^b [f^2(x) + g^2(x)] \, dx.$$

3.19 Use the Cauchy–Schwarz inequality with the two functions $f\sqrt{h}$ and $g\sqrt{h}$.

3.20

(a) Put $a_i = \sqrt{c_i}$ and $b_i = 1/\sqrt{c_i}$ into Cauchy–Schwarz. Equality holds iff $c_1 = \cdots = c_n$.

(b) The given inequality can be rewritten as

$$(\sqrt{a}\sqrt{c} + \sqrt{b}\sqrt{d})^2 \le (a+b)(c+d)$$

and identified as Cauchy–Schwarz.

(c) Apply Cauchy–Schwarz as

$$a\cdot\sqrt{a^2+c^2} + \sqrt{b^2+c^2}\cdot b \le \sqrt{a^2+b^2+c^2}\sqrt{a^2+c^2+b^2}.$$

(d) Apply Cauchy–Schwarz as

$$\frac{1}{\sqrt{b}}\cdot\frac{1}{\sqrt{c}} + \frac{1}{\sqrt{c}}\cdot\frac{1}{\sqrt{a}} + \frac{1}{\sqrt{a}}\cdot\frac{1}{\sqrt{b}} \le \sqrt{\frac{1}{b}+\frac{1}{c}+\frac{1}{a}}\cdot\sqrt{\frac{1}{c}+\frac{1}{a}+\frac{1}{b}}$$

(e) Apply Cauchy–Schwarz as

$$\sqrt{c}\sqrt{a-c} + \sqrt{b-c}\sqrt{c} \le \sqrt{c+(b-c)}\sqrt{(a-c)+c} = \sqrt{ab}.$$

3.21

(a)

$$\left(\sum_{k=1}^n a_k b_k\right)^2 = \left(\sum_{k=1}^n k^{1/2}a_k\cdot\frac{b_k}{k^{1/2}}\right)^2 \le \sum_{k=1}^n ka_k^2\sum_{k=1}^n\frac{b_k^2}{k}.$$

(b)

$$\left(\sum_{k=1}^n a_k^m\right)^2 = \left(\sum_{k=1}^n a_k^{\frac{m+s}{2}}a_k^{\frac{m-s}{2}}\right)^2 \le \sum_{k=1}^n a_k^{m+s}\sum_{k=1}^n a_k^{m-s}.$$

(c)

$$\left(\sum_{k=1}^n\frac{a_k}{k}\right)^2 = \left(\sum_{k=1}^n k^{3/2}a_k\cdot\frac{1}{k^{5/2}}\right)^2 \le \sum_{k=1}^n k^3a_k^2\sum_{k=1}^n\frac{1}{k^5}.$$

(d)

$$\left(\sum_{k=1}^n a_kb_kc_k\right)^4 = \left(\sum_{k=1}^n(a_kb_k)c_k\right)^4 \le \left(\sum_{k=1}^n(a_kb_k)^2\right)^2\left(\sum_{k=1}^n c_k^2\right)^2 \le \left(\sum_{k=1}^n a_k^4\right)\left(\sum_{k=1}^n b_k^4\right)\left(\sum_{k=1}^n c_k^2\right)^2.$$

(e) Set $b_k = 1$ for all k.

(g)

$$\left(\sum_{k=1}^n a_kb_kc_kd_k\right)^4 = \left[\left(\sum_{k=1}^n(a_kb_k)(c_kd_k)\right)^2\right]^2 \le \left[\sum_{k=1}^n(a_kb_k)^2\sum_{k=1}^n(c_kd_k)^2\right]^2$$

$$= \left(\sum_{k=1}^n a_k^2b_k^2\right)^2\left(\sum_{k=1}^n c_k^2d_k^2\right)^2 \le \sum_{k=1}^n a_k^4\sum_{k=1}^n b_k^4\sum_{k=1}^n c_k^4\sum_{k=1}^n d_k^4.$$

(h)

$$\left(\sum_{k=1}^n a_kb_k\right)^2 = \left(\sum_{k=1}^n\sqrt{a_k}\cdot\sqrt{a_k}\,b_k\right)^2 \le \sum_{k=1}^n a_k\sum_{k=1}^n a_kb_k^2.$$

(i)

$$\left(\sum_{k=1}^{n} b_k\right)^2 = \left(\sum_{k=1}^{n} \sqrt{a_k} \cdot \frac{b_k}{\sqrt{a_k}}\right)^2 \le \sum_{k=1}^{n} a_k \sum_{k=1}^{n} \frac{b_k^2}{a_k}.$$

3.22 Put $f(x) = \sqrt{g(x)} \cdot f(x)/\sqrt{g(x)}$ and use the Cauchy–Schwarz inequality.

3.23 [27] Multiply by 1 and use Cauchy–Schwarz to get

$$\int_a^b |f(x)| \, dx \le (b-a)^{1/2} \left(\int_a^b |f(x)|^2 \, dx\right)^{1/2}.$$

Now let $f(x) = 1/\sqrt{x}$ on $[1, r^2]$ where $r > 1$. Obtain

$$\int_1^{r^2} \frac{dx}{\sqrt{x}} \le (r^2 - 1)^{1/2} \left(\int_1^{r^2} \frac{dx}{x}\right)^{1/2}.$$

which reduces to

$$2(r-1) \le (r^2 - 1)^{1/2}(2\ln r)^{1/2}.$$

Factoring $(r^2 - 1)^{1/2}$ as $(r+1)^{1/2}(r-1)^{1/2}$, canceling a factor of $(r-1)^{1/2}$, and then squaring both sides, we get the desired inequality.

3.25 Inscribe a polygon of N sides in a circle of fixed radius R. With θ_n the central angle subtending the nth side of the polygon, the area of the polygon is A, where

$$A = \sum_{n=1}^{N} \frac{R^2}{2} \sin \theta_n = \frac{NR^2}{2} \frac{1}{N} \sum_{n=1}^{N} \sin \theta_n \le \frac{NR^2}{2} \sin\left(\frac{1}{N} \sum_{n=1}^{N} \theta_n\right).$$

Thus $A \le (NR^2/2)\sin(2\pi/N)$, and equality holds only with all central angles equal.

3.26

(b)   Write $f_n(px_1 + (1-p)x_2) \le p f_n(x_1) + (1-p)f_n(x_2)$ and let $n \to \infty$.

(d) $f''(x) = 1/x > 0$ on $(0, \infty)$.

3.27

(a)   With $f(x) = x^2$ and $\delta_k = 1/n$ for $k = 1, \ldots, n$, Jensen's inequality becomes

$$\left(\frac{1}{n} \sum_{k=1}^{n} x_k\right)^2 \le \sum_{k=1}^{n} \frac{1}{n} x_k^2.$$

Note that the result was one of the consequences of the Cauchy–Schwarz inequality in Problem 3.21.

(b)   Divide through by 2 and exploit the convexity of $x \ln x$.

(c)   Since $f''(x) > 0$, $f(x)$ is convex. Therefore

$$-\ln\left(\sum_{k=1}^{n} \delta_k a_k\right) \le -\sum_{k=1}^{n} \delta_k \ln a_k$$

which can be written as

$$\ln\left(\sum_{k=1}^{n} \delta_k a_k\right) \ge \ln\left(\prod_{k=1}^{n} a_k^{\delta_k}\right).$$

Now use monotonicity of the ln function.

(d)   Use Jensen's inequality with $f(x) = \ln(1 + e^x)$.

3.28 (By now this is old hat.) Let

$$\Delta t = (b-a)/n, \quad t_i = a + i\Delta t, \quad c_i = \frac{p(t_i)}{\sum_{j=1}^{n} p(t_j)}, \quad \text{and} \quad x_i = g(t_i).$$

Apply (3.37) and Lemma 2.1.

4.1

(a)   The statement is equivalent to

$$-d(y,z) \le d(x,y) - d(x,z) \le d(y,z).$$

The left inequality is equivalent to $d(x,z) \le d(x,y) + d(y,z)$, while the right is equivalent to $d(x,y) \le d(x,z) + d(z,y)$. But these are both occurrences of the triangle inequality.

(b)   Use induction to generalize the triangle inequality.

4.2 Write $d(x,y) \le d(x,z) + d(z,u) + d(u,y)$ to get

$$d(x,y) - d(z,u) \le d(x,z) + d(u,y).$$

Swap x with z and y with u to get

$$d(z,u) - d(x,y) \le d(x,z) + d(u,y),$$

hence

$$|d(x,y) - d(z,u)| \le d(x,z) + d(u,y).$$

Assume $x_n \to x$ and $y_n \to y$, and use the inequality to write

$$|d(x_n, y_n) - d(x,y)| \le d(x_n, x) + d(y_n, y) \to 0$$

which shows that

$$\lim_{n \to \infty} d(x_n, y_n) = d(x,y).$$

4.3 Assume x and y are distinct limits for $\{x_n\}$. Then for sufficiently large n,

$$d(x,y) \le d(x, x_n) + d(x_n, y) < \varepsilon/2 + \varepsilon/2 = \varepsilon$$

by the triangle inequality. Because ε is arbitrarily small, we must have $x = y$, a contradiction.

4.4

(a)   Use Minkowski's inequality.

(b)   Verification of the first two metric space requirements is trivial. For the third, use the triangle and Minkowski's inequalities as follows:

$$d(\xi, \eta) = \left(\sum_{i=1}^{\infty} |\xi_i - \zeta_i + \zeta_i - \eta_i|^p \right)^{1/p} \le \left(\sum_{i=1}^{\infty} [|\xi_i - \zeta_i| + |\zeta_i - \eta_i|]^p \right)^{1/p}$$

$$\le \left(\sum_{i=1}^{\infty} |\xi_i - \zeta_i|^p \right)^{1/p} + \left(\sum_{i=1}^{\infty} |\zeta_i - \eta_i|^p \right)^{1/p} = d(\xi, \zeta) + d(\zeta, \eta).$$

4.7 Let x be the vector from C to β, y the vector from C to α. Then the median vector from α to A is $2x-y$, and the median vector from β to B is $2y-x$. Compare the magnitudes of these median vectors by using the fact that the inner product of any vector with itself equals its magnitude squared.

4.8 By the proof of Theorem 4.8 equality holds iff we have

$$|\langle x, y \rangle| = \sqrt{\langle x, x \rangle \langle y, y \rangle} \quad \text{and} \quad \text{Re}\langle x, y \rangle = |\langle x, y \rangle|,$$

i.e., $\langle x, y \rangle$ is real and nonnegative. Hence equality holds iff $x = 0$ or $y = 0$ or else $x = \beta y$ for some β real and nonnegative.

4.9 Consider each $z_j = a_j + i b_j \in \mathbb{C}$ as a point $w_j = \begin{pmatrix} a_j \\ b_j \end{pmatrix} \in \mathbb{R}^2$.

$$\left| \sum_{i=1}^n z_i \right| = \sqrt{\left(\sum_{i=1}^n a_i \right)^2 + \left(\sum_{i=1}^n b_i \right)^2} = \sqrt{\sum_{i=1}^n \|w_i\|^2 + 2 \sum_{1 \le i < j \le n} \langle w_i, w_j \rangle},$$

whereas

$$\sum_{i=1}^n |z_i| = \sum_{i=1}^n \|w_i\|.$$

Squaring both sides, (1.17) is equivalent to

$$\sum_{i=1}^n \|w_i\|^2 + \sum_{1 \le i < j \le n} 2\langle w_i, w_j \rangle \le \left(\sum_{i=1}^n \|w_i\| \right)^2 = \sum_{i=1}^n \|w_i\|^2 + \sum_{1 \le i < j \le n} 2\|w_i\| \|w_j\|.$$

By Cauchy–Schwarz for each i, j,

$$\langle w_i, w_j \rangle \le \|w_i\| \|w_j\|$$

hence the inequality is established. Furthermore, equality holds in the sum iff

$$\langle w_i, w_j \rangle = |\langle w_i, w_j \rangle| = \|w_i\| \|w_j\| \quad \text{for all } i < j.$$

Hence equality holds iff for each $i < j$ we have $w_i = \beta_{ij} w_j$ for some constant $\beta_{ij} > 0$, i.e., $\arg z_i = \arg z_j$.

4.10 Write down the norm of Af and use the Schwarz inequality.

4.11 For $m > n$ we have

$$d(x_m, x_n) \le d(x_n, x_{n+1}) + d(x_{n+1}, x_{n+2}) + \cdots + d(x_{m-1}, x_m) \le 2^{-n} + 2^{-(n+1)} + \cdots + 2^{-(m-1)}$$

$$= 2^{-n}[1 + 2^{-1} + \cdots + 2^{-(m-n-1)}] < 2^{-n} \sum_{k=0}^{\infty} 2^{-k} = 2^{-n} \cdot 2 \to 0 \text{ as } n \to \infty.$$

4.12 Both parts are applications of the triangle inequality for the norm.

(a) $\|(x_m + y_m) - (x + y)\| = \|(x_m - x) + (y_m - y)\| \le \|x_m - x\| + \|y_m - y\| \to 0.$

(b) $\|\lambda_m x_m - \lambda x\| = \|\lambda_m (x_m - x) + (\lambda_m - \lambda)x\| \le |\lambda_m| \|x_m - x\| + |\lambda_m - \lambda| \|x\| \to 0.$

5.1 If necessary, multiply the integrand by 1 before applying the Cauchy–Schwarz inequality.

(a) $I_1 < \sqrt{5/4}.$

(b) $I_2 < \pi.$

(c) $I_3 < \sqrt{2\pi}.$

5.2

(a) $I \le 32.58$ (actual value close to 28).

(b) Start with

$$\int_0^t \left(\frac{1}{\sqrt{1 - x^2}} \right)^2 dx > \frac{1}{t} \left(\int_0^t \frac{dx}{\sqrt{1 - x^2}} \right)^2.$$

5.3 We have $|z^2 + a^2| \geq |a^2 - |z|^2| = |a^2 - b^2|$ and the length of C is

$$L = \int_0^{\theta_0} b \, d\theta = b\theta_0 \, ,$$

so

$$|I| \leq \frac{b\theta_0}{|a^2 - b^2|} \, .$$

5.4 If $u(x)$ and every $u_n(x)$ are integrable on $[a, b]$, then

$$\left| \int_a^b u_n(x) \, dx - \int_a^b u(x) \, dx \right| = \left| \int_a^b [u_n(x) - u(x)] \, dx \right| \leq \int_a^b |u_n(x) - u(x)| \, dx \, .$$

Hence if $\{u_n(x)\}$ converges uniformly to $u(x)$ on $[a, b]$, then for $\varepsilon > 0$ there exists N such that

$$\left| \int_a^b u_n(x) \, dx - \int_a^b u(x) \, dx \right| < (b - a)\varepsilon \quad \text{whenever } n > N \, .$$

5.5

(a) Given $\varepsilon > 0$, take N so large that $n > N$ implies $|u(x) - u_n(x)| < [\varepsilon/(b-a)]^{1/2}$ for all $x \in [a, b]$. Then for $n > N$,

$$\int_a^b [u(x) - u_n(x)]^2 \, dx < \int_a^b \varepsilon/(b - a) \, dx = \varepsilon \, .$$

(b) Use the inequality

$$\left| \left(\int_a^b u^2 \, dx \right)^{1/2} - \left(\int_a^b u_n^2 \, dx \right)^{1/2} \right| \leq \left(\int_a^b (u - u_n)^2 \, dx \right)^{1/2} \, .$$

5.6 Convert the following pseudo-code to your favorite language:

```
tol=.5E-3; a=0; b=1; f(x)=e^(x^2); n=2; h=(b-a)/n; ends=f(a)+f(b);
evens=0; odds=f(a+h); aold=(h/3)(ends+4*odds+2*evens);
DoLoop
   n=2*n;
   h=h/2;
   evens=evens+odds;
   odds=f(a+h)+f(a+3h)+...+f(a+(n-1)h);
   anew=(h/3)(ends+4*odds+2*evens)
   if |anew-aold|<=tol*|anew| then exit
     else
   aold=anew;
End Doloop
Print anew
```

5.7 Some helpful formulas are [31]

$$\Gamma(1/2) = \sqrt{\pi} \, , \qquad \Gamma(n) = (n - 1)! \, , \qquad \Gamma(n + \tfrac{1}{2}) = \frac{\sqrt{\pi}}{2^n} (2n - 1)!! \, .$$

Using these the given inequality can be put into the form

$$2 \leq \frac{2^n n!}{(2n - 1)!!} \leq 2^n$$

which is easily established by induction.

5.8 $t > 1 \implies e^{-xt}/t^n > e^{-xt}/t^{n+1}$.

5.9 We have

$$\int_x^\infty \frac{e^{-t}}{t}\,dt < \int_x^\infty \frac{e^{-t}}{x}\,dt = \frac{e^{-x}}{x}.$$

On the other hand, integration by parts gives

$$\int_x^\infty \frac{e^{-t}}{t}\,dt = \frac{e^{-x}}{x} - \int_x^\infty \frac{e^{-t}}{t^2}\,dt > \frac{e^{-x}}{x} - \frac{e^{-x}}{x^2} = \frac{x-1}{x^2}\,e^{-x}.$$

5.10 Use

$$\left(\int_{-\infty}^\infty f(u)f(t+u)\,du\right)^2 \le \int_{-\infty}^\infty f^2(u)\,du \int_{-\infty}^\infty f^2(t+u)\,du = \left(\int_{-\infty}^\infty f^2(u)\,du\right)^2 = (f \star f(0))^2$$

where equality holds iff $t = 0$.

5.11 (See [89].) Starting with

$$|J_n(x)| = \left|\sum_{m=0}^\infty \frac{(-1)^m (x/2)^{2m+n}}{m!(m+n)!}\right|$$

apply the triangle inequality to get

$$|J_n(x)| \le \left|\frac{x}{2}\right|^n \sum_{m=0}^\infty \frac{(x^2/4)^m}{m!(m+n)!} = \frac{1}{n!}\left|\frac{x}{2}\right|^n \sum_{m=0}^\infty \frac{(x^2/4)^m}{m!(n+1)(n+2)\cdots(n+m)}$$

$$\le \frac{1}{n!}\left|\frac{x}{2}\right|^n \sum_{m=0}^\infty \frac{(x^2/4)^m}{m!(n+1)^m} = \frac{1}{n!}\left|\frac{x}{2}\right|^n \exp\left[\frac{(x^2/4)}{n+1}\right] \le \frac{1}{n!}\left|\frac{x}{2}\right|^n \exp\left[\frac{x^2}{4}\right].$$

5.12

(a) Show that erfc$(\sqrt{x}) = 2Q(\sqrt{2x})$. Then use the upper bound for $Q(x)$ that appears in (5.49).

(b) Letting $y = x + d$, we have

$$\int_{\sqrt{y}}^\infty e^{-t^2}\,dt = \int_{\sqrt{x+d}}^\infty e^{-t^2}\,dt = \int_x^\infty \frac{e^{-(u+d)}}{2\sqrt{u+d}}\,du \le e^{-d} \int_x^\infty \frac{e^{-u}}{2\sqrt{u}}\,du = e^{-d} \int_{\sqrt{x}}^\infty e^{-t^2}\,dt.$$

5.13

(c) For $n \ge 1$,

$$a_{n+1} - b_{n+1} = \frac{1}{2}\frac{\sqrt{a_n} - \sqrt{b_n}}{\sqrt{a_n} + \sqrt{b_n}}(a_n - b_n) \le \frac{1}{2}(a_n - b_n)$$

so

$$a_2 - b_2 \le \frac{1}{2}(a_1 - b_1)\,, \quad \ldots\,, \quad a_{n+1} - b_{n+1} \le \frac{1}{2^n}(a_1 - b_1)\,,$$

and now use the squeeze principle on

$$0 \le a_{n+1} - b_{n+1} \le \frac{1}{2^n}(a_1 - b_1)\,.$$

(d) Rewrite the integrand of $T(a,b)$ as

$$\frac{1}{\sin x \cos x\,[(a+b)^2 + (a\cot x - b\tan x)^2]^{1/2}}\,.$$

Sketch $a \cot x - b \tan x$ to see that we may substitute

$$\tan y = \frac{a \cot x - b \tan x}{2b_1} \quad \text{for } -\pi/2 < y < \pi/2 .$$

Then

$$\frac{1}{\cos y} = \frac{[(2b_1)^2 + (a \cot x - b \tan x)^2]^{1/2}}{2b_1} .$$

Now using $b_1^2 = ab$ and a few steps of algebra,

$$\frac{2b_1}{\cos y} = a \cot x + b \tan x .$$

Since also $2b_1 \tan y = a \cot x - b \tan x$, subtraction yields

$$b \tan x = \frac{b_1}{\cos y} - b_1 \tan y .$$

Next, differentiate to get

$$\frac{b \, dx}{\cos^2 x} = \frac{b_1 (\sin y - 1) \, dy}{\cos^2 y} .$$

Multiplication of both sides by $\cot x$ gives

$$\frac{b \, dx}{\sin x \cos x} = \frac{b_1 (\sin y - 1) \, dy}{\cos^2 y \tan x} .$$

Now use

$$\frac{1}{\tan x} = \frac{b \cos y}{b_1 (1 - \sin y)} \quad \text{to get} \quad \frac{dx}{\sin x \cos x} = -\frac{dy}{\cos y} ,$$

hence

$$T(a, b) = \frac{1}{2} \int_{-\pi/2}^{\pi/2} \frac{\sec y \, dy}{\sqrt{a_1^2 + b_1^2 \tan^2 y}} = \int_0^{\pi/2} \frac{dy}{\sqrt{a_1^2 \cos^2 y + b_1^2 \sin^2 y}} .$$

(e)

$$b_n = \sqrt{b_n^2 \cos^2 x + b_n^2 \sin^2 x} \le \sqrt{a_n^2 \cos^2 x + b_n^2 \sin^2 x} \le \sqrt{a_n^2 \cos^2 x + a_n^2 \sin^2 x} = a_n .$$

Use the squeeze principle.

(f) $T(a, b) = T(a_1, b_1)$. By induction, we have $T(a, b) = T(a_n, b_n)$ for all n. Since the sequence

$$\left\{ (a_n^2 \cos^2 x + b_n^2 \sin^2 x)^{1/2} \right\}$$

converges uniformly we may pass the limit inside the integral:

$$T(a, b) = T(a_n, b_n) = \int_0^{\pi/2} \frac{dx}{\sqrt{a_n^2 \cos^2 x + b_n^2 \sin^2 x}} \to \int_0^{\pi/2} \frac{dx}{AG(a, b)} = \frac{\pi/2}{AG(a, b)} .$$

5.14 Put $\phi = \psi = \Phi$ in Green's formula (5.26) to get

$$\oint_S \Phi \frac{\partial \Phi}{\partial n} \, dS = \int_V \nabla \Phi \cdot \nabla \Phi \, dV + \int_V \Phi \nabla^2 \Phi \, dV = \int_V |\nabla \Phi|^2 \, dV \ge 0 .$$

Here we have used the fact that Laplace's equation holds in V.

5.15 Let the first body carry charge Q_1 at surface potential Φ_1, the second body Q_2 at potential Φ_2; the individual capacitances are then

$$C_1 = Q_1/\Phi_1 \quad \text{and} \quad C_2 = Q_2/\Phi_2 .$$

After the bodies are put in communication the new charges become Q_1' and Q_2', respectively, and the shared potential becomes Φ, where

$$\Phi = Q_1'/C_1 = Q_2'/C_2 \quad \text{and} \quad Q_1' + Q_2' = Q = Q_1 + Q_2 .$$

These equations yield $\Phi = Q/(C_1 + C_2)$ so that the overall capacitance is $C_1 + C_2$. Now it is straightforward to show that the energy stored by any conducting body is given by $W = Q^2/2C$, where Q is its charge and C is its capacitance. Assuming Q_1 and Q_2 are both positive, the AM–GM inequality gives

$$2Q_1 Q_2 C_1 C_2 \le Q_2^2 C_1^2 + Q_1^2 C_2^2 ,$$

and some algebraic manipulation yields

$$(Q_1 + Q_2)^2/(C_1 + C_2) \le Q_1^2/C_1 + Q_2^2/C_2$$

as desired. In case $Q_2 = 0$, the desired inequality is

$$Q_1^2/(C_1 + C_2) \le Q_1^2/C_1 ,$$

which is obvious.

5.17 The condition for equality in the Cauchy–Schwarz inequality (5.38) is

$$\frac{df}{dt} = K_1 t f(t) .$$

This yields

$$\frac{1}{f}\frac{df}{dt} = \frac{d}{dt}[\ln f(t)] = K_1 t$$

so that

$$\ln f(t) = \frac{K_1 t^2}{2} .$$

5.18 Write

$$Q(x) = \frac{1}{\sqrt{2\pi}} \int_x^\infty e^{-t^2/2}\, dt = -\frac{1}{\sqrt{2\pi}} \int_x^\infty \frac{1}{t}\frac{d}{dt} e^{-t^2/2}\, dt$$

and integrate by parts to get

$$Q(x) = \frac{1}{\sqrt{2\pi}x} e^{-x^2/2} - \frac{1}{\sqrt{2\pi}} \int_x^\infty \frac{e^{-t^2/2}}{t^2}\, dt .$$

But

$$\frac{1}{\sqrt{2\pi}} \int_x^\infty \frac{e^{-t^2/2}}{t^2}\, dt < \frac{1}{x^2}\frac{1}{\sqrt{2\pi}} \int_x^\infty e^{-t^2/2}\, dt = \frac{1}{x^2} Q(x) .$$

Hence

$$Q(x) > \frac{1}{\sqrt{2\pi}x} e^{-x^2/2} - \frac{1}{x^2} Q(x) .$$

Now solve for $Q(x)$.

5.19 The Q-function is

$$Q(t) = \int_x^\infty \frac{1}{\sqrt{2\pi}} e^{-\tau^2/2}\, d\tau .$$

The key observation: $[Q(t)]^2$ equals the probability that a pair of independent standard-normal Gaussian random variables (X, Y) will fall within the region

$$R_1 = \{(X, Y): X \geq t, Y \geq t\} .$$

This is certainly exceeded by the probability that (X, Y) will fall within the region

$$R_2 = \{(X, Y): X^2 + Y^2 \geq 2t^2\} ,$$

and this latter probability can be evaluated using a change to polar coordinates:

$$\int_{R_2} \frac{1}{2\pi} e^{-(u^2+v^2)/2} \, du \, dv = \frac{1}{2\pi} \int_0^{\pi/2} \int_{\sqrt{2t}}^{\infty} e^{-\rho^2/2} \rho \, d\rho \, d\phi = \frac{1}{4} \int_{\sqrt{2t}}^{\infty} e^{-\rho^2/2} \rho \, d\rho = -\frac{1}{4} e^{-\rho^2/2} \Big|_{\sqrt{2t}}^{\infty} = \frac{1}{4} e^{-t^2} .$$

So we have

$$[Q(t)]^2 \leq \frac{1}{4} e^{-t^2} ,$$

and this gives the quoted bound.

5.20 The frequency function of X is

$$P(X = n) = \frac{\lambda^n}{n!} e^{-\lambda} \qquad (n = 0, 1, 2, \ldots) .$$

We have

$$E[e^{sX}] = \sum_{n=0}^{\infty} e^{sn} \frac{\lambda^n}{n!} e^{-\lambda} = e^{-\lambda} \sum_{n=0}^{\infty} \frac{(e^s \lambda)^n}{n!} = e^{-\lambda} e^{\lambda e^s} = e^{\lambda(e^s - 1)} .$$

The Chernoff bound is

$$P(X \geq k) \leq \inf_{s>0} e^{-sk} e^{\lambda(e^s-1)} = \inf_{s>0} e^{\lambda(e^s-1)-sk}$$

and the expression on the right is minimized when $s = \ln(k/\lambda)$.

5.21 Substitute $\omega = y - 1$ so that

$$\omega' = 2\omega + 2 , \quad \omega(0) = 0 .$$

For this shifted differential equation,

$$\phi_0(x) = 0 ,$$

$$\phi_1(x) = \int_0^x f(t, \phi_0(t)) \, dt = \int_0^x (2\phi_0(t) + 2) \, dt = 2x ,$$

$$\vdots$$

$$\phi_n(x) = \frac{(2x)^n}{n!} + \cdots + \frac{(2x)^2}{2} + 2x .$$

Recognize $\phi_n(x)$ as the Taylor polynomial of degree n for $e^{2x} - 1$, hence

$$\phi_n(x) \to e^{2x} - 1 .$$

Therefore $y = e^{2x}$ solves the original differential equation.

5.22 In general we want to choose a neighborhood $\overline{U}_1$ of ξ so that

$$\|G'(x)\| \leq \alpha < 1 \quad \text{for all } x \in \overline{U}_1 .$$

This insures that G is a contraction. For the case $n = 1$,

$$G(x) = x - \frac{F(x)}{F'(x)} \quad \text{and} \quad G'(x) = \frac{F(x)F''(x)}{[F'(x)]^2} \; .$$

So if an initial guess x is sufficiently close to ξ so that

$$\left| \frac{F(x)F''(x)}{[F'(x)]^2} \right| \le \alpha < 1 \; ,$$

then convergence is guaranteed (at least in theory).

6.1 The system is equivalent to the integral equation

$$\mathbf{x}(t) = \mathbf{x}_0 + \int_{t_0}^{t} A(s)\mathbf{x}(s)\,ds + \int_{t_0}^{t} \mathbf{f}(s)\,ds$$

from which we get

$$\|\mathbf{x}(t)\| \le \|\mathbf{x}_0\| + \int_{t_0}^{t} \|A(s)\| \, \|\mathbf{x}(s)\| \, ds + \int_{t_0}^{t} \|\mathbf{f}(s)\| \, ds \; .$$

By Gronwall's inequality we have

$$\|\mathbf{x}(t)\| \le \left(\|\mathbf{x}_0\| + \int_{t_0}^{T} \|\mathbf{f}(s)\| \, ds \right) \exp\left[\int_{t_0}^{t} \|A(s)\| \, ds \right] \; .$$

6.2 Reduce the equation to the system of Problem 6.1 by introducing $\mathbf{x} = (x_1, \ldots, x_n)^T$, where

$$x_1(t) = y(t), \ldots, x_n(t) = y^{n-1}(t) \; , \quad \mathbf{x}(a) = (y_0, \ldots, y_{n-1})^T \; , \quad \mathbf{f}(t) = (0, \ldots, f(t))^T \; ,$$

and the corresponding matrix $A(t)$.

6.3 Repeat everything with summation up to 2.

6.4 Using the calculus of variations it can be shown that we can find a solution to the boundary value problem by minimizing the functional

$$\frac{1}{2} \int_{a}^{b} [p(x)(y'(x))^2 + q(x)y^2(x)]\,dx - \int_{a}^{b} f(x)y(x)\,dx$$

over the set $C_0^{(2)}$ of functions twice continuously differentiable on $[a, b]$ and vanishing at points a and b. Let $\phi_1, \ldots, \phi_n \in C_0^{(2)}$ be linearly independent. Ritz's approximation is sought in the form

$$y_n = c_1\phi_1 + \cdots + c_n\phi_n \; .$$

Introduce

$$\langle y, z \rangle = \int_{a}^{b} [p(x)y'(x)z'(x) + q(x)y(x)z(x)]\,dx \; ,$$

which possesses all properties of an inner product on $C_0^{(2)}$, and

$$F(y) = \int_{a}^{b} f(x)y(x)\,dx \; .$$

The Ritz method is to minimize the following function of the real variables $c_1, \ldots, c_n$:

$$W\left(\sum_{k=1}^{n} c_k\phi_k \right) = \frac{1}{2}\left\langle \sum_{k=1}^{n} c_k\phi_k, \sum_{k=1}^{n} c_k\phi_k \right\rangle - F\left(\sum_{k=1}^{n} c_k\phi_k \right).$$

The equations of the method are

$$\frac{\partial}{\partial c_k} W\left(\sum_{k=1}^{n} c_k \phi_k\right) = 0 \qquad (k = 1, \ldots, n)$$

which take the form

$$\left\langle \sum_{k=1}^{n} c_k \phi_k, \phi_m \right\rangle = F(\phi_m) \qquad (m = 1, \ldots, n) .$$

6.5 Integrate by parts in Galerkin's equations and use the notations of Problem 6.4.

7.1

(a)
$$1 + \frac{B}{A} = [1 + \underline{B}/\overline{A}, \ 1 + \overline{B}/\underline{A}] .$$

(b)
$$\frac{AB}{B + C} \rightarrow \frac{A}{1 + C/B} = \left[\frac{\underline{A}}{1 + \overline{C}/\underline{B}}, \ \frac{\overline{A}}{1 + \underline{C}/\overline{B}} \right] .$$

(c)
$$\frac{AB}{CD} = [\underline{A}\,\underline{B}/\overline{C}\,\overline{D}, \ \overline{A}\,\overline{B}/\underline{C}\,\underline{D}] .$$

(d)
$$A + B + C = [\underline{A} + \underline{B} + \underline{C}, \ \overline{A} + \overline{B} + \overline{C}] .$$

(e)
$$\frac{B + C}{A} = \left[\frac{\underline{B} + \underline{C}}{\overline{A}}, \ \frac{\overline{B} + \overline{C}}{\underline{A}} \right] .$$

(f)
$$\frac{A + BC}{D} = \left[\frac{\underline{A} + \underline{B}\,\underline{C}}{\overline{D}}, \ \frac{\overline{A} + \overline{B}\,\overline{C}}{\underline{D}} \right] .$$

(g)
$$C\left(A + \frac{1}{B}\right) = \left[\underline{C}(\underline{A} + 1/\overline{B}), \ \overline{C}(\overline{A} + 1/\underline{B}) \right] .$$

(h)
$$\frac{A}{B} + \frac{C}{D} = [\underline{A}/\overline{B} + \underline{C}/\overline{D}, \ \overline{A}/\underline{B} + \overline{C}/\underline{D}] .$$

(i)
$$\frac{1}{1 + A(1/B + 1/C)} = [(1 + \overline{A}(1/\underline{B} + 1/\underline{C}))^{-1}, \ (1 + \underline{A}(1/\overline{B} + 1/\overline{C}))^{-1}] .$$

7.2 Note that
$$w(X) = \overline{X} - \underline{X} = \max_{a,b \in X} |a - b| \quad \text{and} \quad |X| = \max_{x \in X} |x| .$$

(a)   Write
$$w(XY) = \max_{\substack{x,x' \in X \\ y,y' \in Y}} |xy - x'y'| \quad \text{where} \quad |xy - x'y'| \le |x||y - y'| + |y'||x - x'| .$$

(b)
$$|\alpha X| = \max_{x \in X} |\alpha x| = |\alpha| \max_{x \in X} |x| = |\alpha|\,|X| .$$

(c)
$$w(\alpha X) = \max_{x,x' \in X} |\alpha x - \alpha x'| = |\alpha| \max_{x,x' \in X} |x - x'| = |\alpha|\,w(X) .$$

(d)

$$|X + Y| = \max_{x \in X, y \in Y} |x + y| \le \max_{x \in X} |x| + \max_{y \in Y} |y| = |X| + |Y| .$$

(e)

$$w(1/X) = \max_{x,x' \in X} \left| \frac{1}{x} - \frac{1}{x'} \right| = \max_{x,x' \in X} \frac{|x - x'|}{|x| \, |x'|} \le \max_{x \in X} \frac{1}{|x|} \max_{x' \in X} \frac{1}{|x'|} \max_{x,x' \in X} |x - x'| = |1/X|^2 \, w(X) .$$

(f)

$$|XY| = \max_{c \in XY} |c| = \max_{a \in X, \, b \in Y} |ab| = \max_{a \in X, \, b \in Y} |a| \, |b| = \max_{a \in X} |a| \max_{b \in Y} |b| = |X| \, |Y| .$$

(g) Obvious.

(h)

$$w(X + Y) = w([\underline{X} + \underline{Y}, \overline{X} + \overline{Y}]) = \overline{X} + \overline{Y} - (\underline{X} + \underline{Y}) = \overline{X} - \underline{X} + \overline{Y} - \underline{Y} = w(X) + w(Y) .$$

(i)

$$w(X - Y) = w([\underline{X} - \overline{Y}, \overline{X} - \underline{Y}]) = \overline{X} - \underline{Y} - (\underline{X} - \overline{Y}) = \overline{X} - \underline{X} + \overline{Y} - \underline{Y} = w(X) + w(Y) .$$

(j)

$$w(XY) = \max_{\substack{x,x' \in X \\ y,y' \in Y}} |xy - x'y'| \ge \max_{\substack{x \in X \\ y,y' \in Y}} |xy - xy'| = \max_{\substack{x \in X \\ y,y' \in Y}} |x| \, |y - y'| = |X| w(Y) .$$

(k)   Follows from the previous part and the similar relation $w(XY) \ge |Y| w(X)$.

7.3 We have

$$A(B + C) = \{z = a(b + c) : a \in A, \, b \in B, \, c \in C\}$$

$$\subseteq \{z = ab + \tilde{a}c : a, \tilde{a} \in A, \, b \in B, \, c \in C\}$$

$$= AB + AC .$$

But note that if $a \in \mathbb{R}$, then

$$a(B + C) = \{z = a(b + c) : b \in B, \, c \in C\}$$

$$= \{z = ab + ac : b \in B, \, c \in C\}$$

$$= aB + aC .$$

7.4 We have

$$X_1 = F(X_0) = 1 + \frac{1}{1 + [1, 2]} = 1 + \frac{1}{[2, 3]} = 1 + 1 \cdot [\tfrac{1}{3}, \tfrac{1}{2}] = [\tfrac{4}{3}, \tfrac{3}{2}] ,$$

$$X_2 = F(X_1) = 1 + \frac{1}{1 + [\tfrac{4}{3}, \tfrac{3}{2}]} = 1 + \frac{1}{[\tfrac{7}{3}, \tfrac{5}{2}]} = 1 + 1 \cdot [\tfrac{2}{5}, \tfrac{3}{7}] = [\tfrac{7}{5}, \tfrac{10}{7}] ,$$

$$X_3 = F(X_2) = 1 + \frac{1}{1 + [\tfrac{7}{5}, \tfrac{10}{7}]} = 1 + \frac{1}{[\tfrac{12}{5}, \tfrac{17}{7}]} = 1 + 1 \cdot [\tfrac{7}{17}, \tfrac{5}{12}] = [\tfrac{24}{17}, \tfrac{17}{12}] ,$$

$$X_4 = F(X_3) = 1 + \frac{1}{1 + [\tfrac{24}{17}, \tfrac{17}{12}]} = 1 + \frac{1}{[\tfrac{41}{17}, \tfrac{29}{12}]} = 1 + 1 \cdot [\tfrac{12}{29}, \tfrac{17}{41}] = [\tfrac{41}{29}, \tfrac{58}{41}] .$$

7.6 The result

$$P_1(x) = 1 + [1, 4]x = [1, 1] + [x, 4x] = [1 + x, 1 + 4x]$$

ensures that

$$1 + x \le y(x) \le 1 + 4x \quad \text{for all } x \in [0, \tfrac{1}{4}] .$$

Plot the actual solution $1/(1 - x)$ along with these rather rough bounds on $[0, \tfrac{1}{4}]$. Repeat for $P_2(x)$.

References

1. Abramowitz, M., Stegun, I.: Handbook of Mathematical Functions. Dover, New York (1965)
2. Alexander, N.: Exploring Biomechanics: Animals in Motion. Scientific American Library, New York (1992)
3. Anderson, G., Vamanamurthy, M., Vuorinen, M.: Conformal Invariants, Inequalities, and Quasiconformal Maps. Wiley, New York (1997)
4. Andrews, L.: Special Functions of Mathematics for Engineers. McGraw-Hill, New York (1992)
5. Ballard, W.: Geometry. W.B. Saunders, Philadelphia (1970)
6. Beckenbach, E., Bellman, R.: An Introduction to Inequalities. Random House, New York (1961)
7. Beckenbach, E., Bellman, R.: Inequalities. Springer, Berlin (1961)
8. Blahut, R.: Principles and Practice of Information Theory. Addison-Wesley, Reading (1987)
9. Börjesson, P., Sundberg, C.: Simple approximations of the error function $Q(x)$ for communications applications. IEEE Trans. Commun. **27**(3), 639–643 (1979)
10. Bradis, V.M., Minkovskii, V.L., Kharcheva, A.K.: Lapses in Mathematical Reasoning. Dover, New York (1999)
11. Brogan, W.: Modern Control Theory. Quantum, New York (1974)
12. Bromwich, T.: An Introduction to the Theory of Infinite Series. Macmillan, London (1965)
13. Bullen, P.S.: A Dictionary of Inequalities. Chapman and Hall/CRC, Boca Raton (1998)
14. Chaplygin, S.A.: New Method of Approximate Integration of Differential Equations. Transactions of Central Aero-Gidrodynamic Institute, 132. GTTI, Moscow-Leningrad (1932)
15. Chow, S., J. Hale.: Methods of Bifurcation Theory. Springer, New York (1982)
16. Couch, L.: Digital and Analog Communication Systems. Macmillan, New York (1990)
17. de Alwis, T.: Projectile motion with arbitrary resistance. Coll. Math. J. **26**(5), 361–367 (1995)
18. Dennis, J., Schnabel, R.: Numerical Methods for Unconstrained Optimization and Nonlinear Equations. Prentice Hall, Englewood Cliffs (1983)
19. Dieudonne, J.: Foundations of Modern Analysis. Academic, New York (1960)
20. Duffin, R.: Cost minimization problems treated by geometric means. Oper. Res. **10**(5), 668–675 (1962)
21. Duffin, R.: Dual programs and minimum cost. J. Soc. Ind. Appl. Math. **10**, 119–123 (1962)
22. Duren, P.: Theory of H^p Spaces. Academic, New York (1970)
23. Edwards, C.: Advanced Calculus of Several Variables. Academic, New York (1973)
24. Eggleston, H.: Elementary Real Analysis. Cambridge University Press, Cambridge (1962)
25. Evans, L.C.: Partial Differential Equations. American Mathematical Society, Providence (1998)
26. Feynmann, R.P., Leighton, R.B., Sands, M.: The Feynman Lectures on Physics, vol. 2. Addison-Wesley, Redwood City (1989)

M.J. Cloud et al., *Inequalities: With Applications to Engineering*,
DOI 10.1007/978-3-319-05311-0, © Springer International Publishing AG 2014

27. Garling, D.J.H.: Inequalities: A Journey Into Linear Analysis. Cambridge University Press, Cambridge (2007)

28. Gelfand, I.: Lectures on Linear Algebra. Dover, New York (1989)

29. Gerrish, F.: Pure Mathematics: A University and College Course. Cambridge University Press, Cambridge (1960)

30. Glaister, P.: Does what goes up take the same time to come down? Coll. Math. J. **24**(2), 155–158 (1993)

31. Gradshteyn, I., Ryzhik, I.: Table of Integrals, Series, and Products. Academic, Boston (1994)

32. Hansen, E.R. (ed.): Topics in Interval Analysis. Oxford University Press, Oxford (1969)

33. Hansen, E., Walster, G.W.: Global Optimization Using Interval Analysis, 2nd edn. Marcel Dekker, New York (2004)

34. Hardy, G., Littlewood, J., Polya, G.: Inequalities. Cambridge University Press, Cambridge (1952)

35. Hartman, Ph.: Ordinary Differential Equations, 2nd edn. SIAM, Philadelphia (2002)

36. Herman, J., Kučera, R., Šimša, J.: Equations and Inequalities: Elementary Problems and Theorems in Algebra and Number Theory. Springer, New York (2000)

37. Hobson, E.: The Theory of Functions of a Real Variable and the Theory of Fourier's Series, vol. 1. Dover, New York (1957)

38. Indritz, J.: Methods in Analysis. Macmillan, New York (1963)

39. Jerri, A.: Introduction to Integral Equations with Applications. Dekker, New York (1985)

40. Jordan, D., Smith, P.: Nonlinear Ordinary Differential Equations. Clarendon, Oxford (1987)

41. Knowles, J.: Energy decay estimates for damped oscillators. Int. J. Math. Educ. Sci. Technol. **28**(1) (1997)

42. Kolev, L.V.: Interval Methods for Circuit Analysis. World Scientific, Singapore (1993)

43. Korovkin, P.P.: Inequalities. Mir Publishers, Moscow (1975)

44. Lafrance, P.: Fundamental Concepts in Communication. Prentice Hall, Englewood Cliffs (1990)

45. Landau, E.: Differential and Integral Calculus, 3rd edn. Chelsea, New York (1980)

46. Lange, K.: Applied Probability. Springer, New York (2010)

47. Lebedev, L.P., Cloud M.J., Eremeyev, V.A.: Advanced Engineering Analysis: The Calculus of Variations and Functional Analysis with Applications in Mechanics. World Scientific, Singapore (2012)

48. Lebedev, L.P., Vorovich, I.I., Cloud, M.J.: Functional Analysis in Mechanics, 2nd edn. Springer, New York (2012)

49. Ledermann, W., Vajda, S.: Handbook of Applicable Mathematics, vol. 4: Analysis. Wiley-Blackwell, New York (1982)

50. Lew, J., Frauenthal, J., Keyfitz, N.: On the average distances in a circular disc. In: Mathematical Modeling: Classroom Notes in Applied Mathematics. SIAM, Philadelphia (1987)

51. Lütkepohl, H: Handbook of Matrices. Wiley, Chichester (1996)

52. Manley, R.: Waveform Analysis. Wiley, New York (1945)

53. Marcus, M., Minc, H.: A Survey of Matrix Theory and Matrix Inequalities. Dover, New York (1992)

54. Massey, H.S.W., Kestelman, H.: Ancillary Mathematics. Pitman, London (1964)

55. Meschkowski, H.: Series Expansions for Mathematical Physicists. Interscience, New York (1968)

56. Mitrinovic, D.S.: Analytic Inequalities. Springer, Berlin (1970)

57. Mitrinovic, D.S., Barnes, E.S., Marsh, D.C.B., Radok, J.R.M.: Elementary Inequalities. P. Noordhoff Ltd, Groningen (1964)

58. Moore, R.E.: The automatic analysis and control of error in digital computation based on the use of interval numbers. In: Rall, L.B. (ed.) Error in Digital Computation, vol. 1. Wiley, New York (1965)

59. Moore, R.E.: Interval Analysis. Prentice Hall, Englewood Cliffs (1966)

60. Moore, R.E.: Methods and Applications of Interval Analysis. SIAM, Philadelphia (1979)

61. Moore, R.E., Kearfott, R.B., Cloud, M.J.: Introduction to Interval Analysis. SIAM, Philadelphia (2009)
62. Moore, R.E.: The dawning. Reliab. Comput. **5**(4), 423–424 (1999)
63. Moore, R.E., Cloud, M.J.: Computational Functional Analysis, 2nd edn. Ellis Horwood, Chichester (2007)
64. Moore, R.E.: Interval Arithmetic and Automatic Error Analysis in Digital Computing. Technical Report 25, Applied Mathematics and Statistics Laboratories, Stanford University (November 1962)
65. Mott, T.E.: On the quotient of monotone functions. Am. Math. Mon. **70**, 195–196 (1963)
66. Neumaier, A.: Interval Methods for Systems of Equations. Cambridge University Press, Cambridge (1990)
67. Nevyazhskii, G.L.: Inequalities: Manual for Teachers. Uchpedgiz, Moscow (1947, in Russian)
68. Olejnik, S.N., Potapov, M.K., Pasichenko, P.I.: Equations and Inequalities. Nonstandard Methods of Solution. Drofa, Moscow (2001, in Russian)
69. Olmsted, J.M.H.: Real Variables, an Introduction to the Theory of Functions. Appleton-Century-Crofts, New York (1959)
70. Pachpatte, B.G.: Mathematical Inequalities. Elsevier, Amsterdam (2005)
71. Papoulis, A.: The Fourier Integral and Its Applications. McGraw-Hill, New York (1962)
72. Patel, V.: Numerical Analysis. Saunders College, New York (1994)
73. Potapov, M.K., Alexsandrov, V.V., Pasichenko, P.I.: Algebra and Analysis of Elementary Functions. Nauka, Moscow (1981, in Russian)
74. Polya, G., Szegö, G.: Isoperimetric Inequalities in Mathematical Physics. Annals of Mathematics Studies, vol. 27. Princeton University Press, Princeton (1951)
75. Protter, M.: Maximum Principles in Differential Equations. Springer, New York (1984)
76. Ptak, V.: The Kantorovich inequality. Am. Math. Mon. **102**(9), 820–821 (1995)
77. Rektorys, K.: Variational Methods in Mathematics, Science, and Engineering. D. Reidel, Boston (1980)
78. Rogosinski, W.: Volume and Integral. Interscience, New York (1952)
79. Rothwell, E.J., Cloud, M.J.: Automatic error analysis using intervals. IEEE Trans. Educ. **55**(1), 9–15 (2012)
80. http://www.ti3.tu-harburg.de/~rump/intlab/. Accessed 10 Jan 2014
81. Sedrakyan, N.M., Avoyan, A.M.: Inequalities: Methods of Proof. Fizmatlit, Moscow (2002, in Russian)
82. Shannon, C.: The Mathematical Theory of Communication. University of Illinois Press, Urbana (1964)
83. Sivashinski, I.: Inequalities in Problems. Nauka, Moscow (1967, in Russian)
84. Stoer, J., Bulirsch, R.: Introduction to Numerical Analysis. Springer, New York (1980)
85. Stratton, J.: Electromagnetic Theory. McGraw-Hill, New York (1941)
86. Stromberg, K.: An Introduction to Classical Real Analysis. Wadsworth International, Belmont (1981)
87. Temme, N.: Special Functions: An Introduction to the Classical Functions of Mathematical Physics. Wiley, New York (1996)
88. http://www.cs.utep.edu/interval-comp/. Accessed 10 Jan 2014
89. Watson, G.: A Treatise on the Theory of Bessel Functions. Cambridge University Press, Cambridge (1944)
90. Weinberger, H.: A First Course in Partial Differential Equations with Complex Variables and Transform Methods. Wiley, New York (1965)

Index

Printed in the United States
By Bookmasters